Visible Light Communications

Vehicular applications

IOP Series in Emerging Technologies in Optics and Photonics

Series Editor

R Barry Johnson a Senior Research Professor at Alabama A&M University, has been involved for over 50 years in lens design, optical systems design, electro-optical systems engineering, and photonics. He has been a faculty member at three academic institutions engaged in optics education and research, employed by a number of companies, and provided consulting services.

Dr Johnson is an IOP Fellow, SPIE Fellow and Life Member, OSA Fellow, and was the 1987 President of SPIE. He serves on the editorial board of Infrared Physics & Technology and Advances in Optical Technologies. Dr Johnson has been awarded many patents, has published numerous papers and several books and book chapters, and was awarded the 2012 OSA/SPIE Joseph W Goodman Book Writing Award for Lens Design Fundamentals, Second Edition. He is a perennial co-chair of the annual SPIE Current Developments in Lens Design and Optical Engineering Conference.

Foreword

Until the 1960s, the field of optics was primarily concentrated in the classical areas of photography, cameras, binoculars, telescopes, spectrometers, colorimeters, radiometers, etc. In the late 1960s, optics began to blossom with the advent of new types of infrared detectors, liquid crystal displays (LCD), light emitting diodes (LED), charge coupled devices (CCD), lasers, holography, fiber optics, new optical materials, advances in optical and mechanical fabrication, new optical design programs, and many more technologies. With the development of the LED, LCD, CCD and other electo-optical devices, the term 'photonics' came into vogue in the 1980s to describe the science of using light in development of new technologies and the performance of a myriad of applications. Today, optics and photonics are truly pervasive throughout society and new technologies are continuing to emerge. The objective of this series is to provide students, researchers, and those who enjoy self-teaching with a wide-ranging collection of books that each focus on a relevant topic in technologies and application of optics and photonics. These books will provide knowledge to prepare the reader to be better able to participate in these exciting areas now and in the future. The title of this series is Emerging Technologies in Optics and Photonics where 'emerging' is taken to mean 'coming into existence,' 'coming into maturity,' and 'coming into prominence.' IOP Publishing and I hope that you find this Series of significant value to you and your career.

Visible Light Communications

Vehicular applications

Xavier Fernando
Ryerson Communications Lab, Canada

Hasan Farahneh
University of Jordan, Jordan

IOP Publishing, Bristol, UK

ISBN 978-0-7503-2284-3 (ebook)
ISBN 978-0-7503-2282-9 (print)
ISBN 978-0-7503-2283-6 (mobi)

DOI 10.1088/978-0-7503-2284-3

Version: 20191101

IOP ebooks

British Library Cataloguing-in-Publication Data: A catalogue record for this book is available from the British Library.

Published by IOP Publishing, wholly owned by The Institute of Physics, London

IOP Publishing, Temple Circus, Temple Way, Bristol, BS1 6HG, UK

US Office: IOP Publishing, Inc., 190 North Independence Mall West, Suite 601, Philadelphia, PA 19106, USA

To Osama, Othman, and Zina Hasan Farahneh

Contents

Preface

The term visible light communication (VLC) refers to a subset of optical wireless communication (OWC). While OWC refers to a communication scheme that uses light waves in general (visible or invisible infrared) as the medium to carry information, the VLC systems exclusively use the visible spectrum (430–770 THz) for communication.

Due to ever increasing scarcity and overcrowding in the radio frequency (RF) spectrum, there is interest in alternative means for wireless communications. OWC systems have been developed in the past few decades, mainly in indoor applications. However, due to the rapid growth of the high power light emitting diode (LED) in the visible spectrum for lighting applications, the interest in VLC has grown rapidly in recent times. VLC is considered an attractive alternative to RF communication due many reasons:

1. High-power, white-light LEDs can be simultaneously used for lighting and communication applications due to their fast switching capabilities.
2. Optical wireless spectrum is license free. Anyone can set-up an OWC system without getting approval from regulating bodies such as the Federal Communications Commission.
3. OWC is both immune to and will not cause electromagnetic interference to other RF communication systems.
4. OWC systems are relatively cost-effective and simple to implement.

Therefore, the application of OWC, especially VLC in automotive (and in-building) applications as an alternative, or more accurately as a complementary solution to the traditional RF-based communications, has created widespread interest. This book investigates the suitability of VLC in such scenarios, especially in vehicular environments.

There is widespread interest in autonomous vehicles (AV)s and intelligent transportation systems (ITS) in the transportation industry. While ITS is aimed towards increasing road safety and for improving traffic efficiency, AVs will transform the way we live and move around. Vehicle-to-anything (V2X) communications are essential for realizing both AVs and ITS. Despite several RF based V2X wireless communications standards investigated for this purpose, there is significant interest in VLC based solutions. VLC is an excellent choice to provide short range line-of-sight (LOS) communication among vehicles and between vehicles and roadside units.

This book covers VLC-based V2X communication considering both non-line-of-sight (NLOS) and LOS signal reception scenarios. The book includes six chapters that cover the important aspects of V2X–VLC. Following this preface, chapter 1 presents a general introduction to VLC. Chapter 2 presents a (2 × 2) MIMO channel for V2V–VLC links by considering both LOS and NLOS propagation scenarios[1].

[1] Note, a pair of vehicle headlights or taillights equipped LED emitters and two photodiode (PD) based VLC receivers can be modeled as a 2 × 2 multiple-inputmultiple-output (MIMO) communication link.

The effects of intermediate reflectors are analyzed by considering infinitesimal Lambertian reflecting elements. Then, the channel impulse response (CIR) is obtained by combining received signal from these reflectors for both single and multiple reflections. The obtained CIR is the backbone for obtaining key statistical properties of the channel such as the received power profile, root mean squared (RMS) delay spread and, coherence bandwidth. These are obtained for vehicles moving in the same direction as well as in opposite directions.

In chapter 3, an advanced modulation scheme, optical-orthogonal frequency division multiplexing (O-OFDM) with adaptive modulation is presented. Note, since the optical wireless (OW) communications environment is a *unipolar* environment, the negative amplitude signal of the bipolar OFDM scheme has to be removed somehow. For this two schemes are used: direct current optical orthogonal frequency division multiplexing (DCO-OFDM) and asymmetrically clipped optical orthogonal frequency division multiplexing (ACO-OFDM). DCO-OFDM is a form of O-OFDM with an appropriate DC bias added to lift-up the entire waveform. In the ACO-OFDM technique, the modulated data is assigned to the odd indexed subcarriers, while the even indexed subcarriers are set to zero. This also removes the negative portion. Numerical simulations show data rates up to 50 Mbps with low BER can be achieved with these algorithms under good signal conditions.

Note, the outdoor VLC channel has frequency selective nature. Hence, in chapter 4, we investigate precoding and equalizing scheme for the V2V–VLC system considering flickering/dimming control. We present the precoding matrix, as well as the equalization matrix for Zero forcing (ZF), maximum likelihood (ML), and minimum mean-square error (MMSE) receiving schemes. The MMSE scheme shows a noticeable improvement over other schemes.

One major issue with the directional nature of light wave propagation is shadowing that impairs communications. Chapter 5 deals with the shadowing effect on the V2V–VLC system. We propose a method to take advantage of the optical diffraction phenomenon to overcome shadowing effect by employing a receiver with a wide field of view (FOV). The shadowing effect for visible light was modelled by a bimodal distribution and the probability of error for different depths of shadowing is derived.

A major problem of outdoor VLC communication systems is sunlight. Solar irradiance drastically affects the V2V–VLC system during daytime. This in-band noise cannot be easily filtered out using optical or electrical bandpass filters. Different denoising schemes are investigated chapter 6. The performance of the system is investigated in terms of SNR, BER, and data rate under three different scenarios. We have proposed filtering the solar irradiance by using a differential receiver that works with two LEDs each with slightly different wavelengths. Also, we propose denoising the solar irradiance in a VLC link using a machine learning (MLE) approach. We propose a k-nearest neighbour (kNN) algorithm-MLE based adaptive filter to combat the effect of solar irradiance. A supervised algorithm was used in the analysis. Both denoising schemes are discussed in detail.

Author biographies

Xavier Fernando

Xavier Fernando is a Professor and Director of Ryerson Communications Lab. He was an IEEE Distinguished Lecturer and delivered over 50 invited lectures worldwide. He has (co-)authored over 200 research articles and holds three patents. He has mono graphed a widely selling book on radio over fiber systems. He was a member in the IEEE COMSOC Education Board Working Group on Wireless Communications, 2010–2012. He is the General Chair of IEEE International Humanitarian Technology Conference, 2017. He was the General Chair for IEEE Canadian Conference on Electrical and Computer Engineering (CCECE) 2014. He and his students have won several awards and prizes including the First prize at the Humanitarian Initiatives Workshop of CCECE 2014; Second Prize at the 2014 IHTC Conference in Montreal; Best Paper Award at the International Conference of Smart Grid Engineering (SEGE 2014), UOIT, Canada, 2014; IEEE Microwave Theory and Techniques Society Prize in 2010, Sarnoff Symposium prize in 2009, Opto-Canada best poster prize in 2003 and CCECE best paper prize in 2001. Xavier Fernando earned his PhD from the University of Calgary, Alberta in 2001 in affiliation with TRLabs.

Hasan Farahneh

Hasan Farahneh received the BS degree from Al-Yarmouk University, Irbid, Jordan in 1985, the MSc degree from the University of Jordan, Amman, Jordan in 2004, and PhD degree from Ryerson University Toronto-Canada in 2018. In September 2018, he joined the Electrical Engineering Department at the University of Jordan as a teaching staff member. His research interest includes wireless and visible light communication. His current research is focused on vehicular communication using visible light spectrum.

List of acronyms

ACO-OFDM	Asymmetrically clipped optical orthogonal frequency division multiplexing
A/D	Analog-to-digital
APD	Avalanche photodiode
AV	Autonomous vehicles
AWGN	Additive white Gaussian noise
BER	Bit error rate
BPF	Band pass filter
BPSK	Binary phase shift keying
BRDF	Bidirectional reflectance distribution function
CDMA	Code-division multiple access
CIR	Channel impulse response
CP	Cyclic prefix
CMOS	Complementary metal-oxide semiconductor
CSI	Channel state information
D/A	Digital-to-analog
DCO-OFDM	Direct current optical orthogonal frequency division multiplexing
DFT	Discrete Fourier transform
DMT	Discrete multi-tone modulation
DSSS	Discrete sequence spread modulation
FFT	Fast Fourier transform
FOV	Field of view
FDM	Frequency-division multiplexing
IDFT	Inverse discrete Fourier transform
IF	Intermediate frequency
IFFT	Inverse fast Fourier transform
IM/DD	Intensity modulation and direct detection
IoT	Internet of Things
IR	Infrared
ISI	Inter symbol interference
ITS	Intelligent transportation systems
IVC	Inter-vehicle communication
I2V	Infrastructure-to-vehicle
kNN	k-nearest neighbour
LED	Light emitting diode
LEDs	Light emitting diodes
LOS	Line-of-sight
LPF	Low pass filter
MIMO	Multiple-input–multiple-output
ML	Maximum likelihood
MLE	Machine learning
MMSE	Minimum mean-square error
MSE	Mean squared error
NLOS	Non-line-of-sight
OCI	Optical communication image sensor
OFDM	Orthogonal frequency division multiplexing
O-OFDM	Optical-orthogonal frequency division multiplexing

OOK	ON-OFF-keying
OW	Optical wireless
OWC	Optical wireless communication
PAM	Pulse amplitude modulation
PD	Photodiode
PDF	Probability density function
PDs	Photodiodes
PIN	Positive intrinsic negative photodiodes
PPM	Pulse position modulation
PSD	Power spectral density
PWM	Pulse width modulation
P/S	Parallel-to-serial
QAM	Quadrature amplitude modulation
QoS	Quality of service
RF	Radio frequency
RMS	Root mean squared
RSU	Road side units
RZ	Return to zero
SISO	Single-input single-output
SNR	Signal-to-noise-ratio
S/P	Serial-to-parallel
STBC	Space-time block coding
SVD	Singular value decomposition
TDE	Time domain equalization
VLC	Visible light communication
V2I	Vehicle-to-infrastructure
V2V	Vehicle-to-vehicle
V2X	Vehicle-to-anything
WDM	Wave division multiplexing
ZF	Zero forcing

List of symbols

A_r	The effective area of the photo detector
A_q	The area of the reflector q
α	The transmitter horizontal angle in relation to the transmitter axis
β	The transmitter vertical angle in relation to the transmitter axis
B_n	Equivalent noise bandwidth
c	The speed of light
$\otimes$	The convolution sign
$d\omega$	The solid angle measured by steradian (sr)
dP_r	The received optical power from any single reflected
δ	The impulse function
E_i	The irradiance incident onto the surface
η	Dimming level
F	Excess noise
F_n	Noise figure of the photodiode
$g(\theta_{\mathrm{nm}})$	The concentrator gain
G	Photodiode gain
γ	The receiver responsivity
h_i	The hight of the photo detector
$h(t)$	The channel impulse response
$I(\alpha, \beta)$	Luminous intensity measured by candela (cd)
I_o	Photo current produced due to the received optical power
I_s	Photo current produced due to sunlight
I_{bg}	Receiver background noise current
I_{dg}	Bulk dark current
I_{ds}	Surface dark current
I_p	The average photocurrent generated at the receiver
K_B	Boltzmann's constant
l	The Lambertian order of the source
L_r	The irradiance scattering from the surface
L_Φ	The illuminance
λ	Light wavelength
$\mu_{0,1}$	The mean values of the Gaussian function
$n(t)$	The noise signal
P_e	Probability of error
Φ	Luminance flux
ϕ_{nm}	The emitting angle from source m to receiver n
ϕ_{inc}	The incident azimuth angle
ϕ_{ref}	The reflected azimuth angle
P_{inc}	The incident optical power
P_r	The received optical power
P_T	The total power
Ψ	The field of view of the receiver
q_e	Electron charge
R_L	Load resistance
ρ	The reflection coefficient of the reflector
$R(\phi)$	The reflected radiant intensity
R_{nm}	The distance between source m and receiver n

θ_{nm}	The incident angle from source *m* to receiver *n*
θ_{inc}	The incident polar angle
θ_{ref}	The reflected polar angle
σ^2	The variance of the signal
T_k	Temperature in kelvin
$T_s(\theta_{\mathrm{nm}})$	The signal transmission coefficient of an optical filter
τ	The time delay
ζ	Biasing ratio

Chapter 1

Introduction

1.1 Optical wireless communication

Optical wireless communications (OWC) is a common name for wireless transmission of information using optical signal or light waves (without an optical fiber). OWC includes both infrared (IR) and visible light. Visible light communication (VLC) is a fast emerging subset of OWC which uses the visible spectrum of the light[1]. The OWC technology has notable potential as a wireless solution that offers greater bandwidth, energy efficiency, security, and data density, while not being subjected to or contributing to electromagnetic interference (EMI) below 3 THz. TV Remote (TVR) deployment is the best ongoing success story of OWC.

Note that VLC systems may serve a dual purpose, simultaneous illumination and transmission of data. This fact has been attracting much attention recently due to the widespread deployment of solid state lights (or LEDs) in buildings and vehicles. Note, LEDs have the ability to turn on and off (or to change their output power) rapidly in response to the change in driving current. This enables them to be directly modulated with a digital (or analog) information signal. Such driving and modulating circuits of LEDs are much simpler and cost effective compared to similar circuits for other types of lights.

This direct modulation results in simple intensity modulation and direct detection (IM/DD) links. Here only the signal intensity is modulated and there is no phase information transmitted. The direct detection receiver detects only the received optical power. Hence the communication signal values in IM/DD systems are real-valued and *unipolar* (non-negative). Hence, popular advanced modulations and signaling schemes of RF wireless communications such as orthogonal frequency division multiplexing (OFDM) and code-division multiple access (CDMA) and their

[1] Except for falling within the visible spectrum range, most characteristics of VLC are the same as that of OWC and in this book we use OWC as the candidate to describe these general characteristics.

doi:10.1088/978-0-7503-2284-3ch1

hybrids cannot be directly used in unipolar IM/DD environments. They need appropriate modifications to avoid the negative portion of the waveform[2].

While RF-based communication systems with limited spectrum are overcrowded with a large number of Internet of Things (IoT) nodes and users, the OWC systems provide an attractive alternative. In addition to its ubiquitous nature, OWC offers huge bandwidth that does not need licensing, hence is available free of charge. OWC inherently has directional propagation properties and typically covers a short range. These characteristics, while perceived as drawbacks by some, drastically decrease potential interference from other communication links and eavesdropping in OWC environments. In addition, OWC does not cause and is also immune to electromagnetic interference from other RF communication systems. These characteristics make it ideal solution for electromagnetically sensitive environments such as in aircraft, in hospitals and close to power cables. Visible light also has excellent propagation characteristics under water, making it ideal for underwater communications.

On the other hand, VLC also has a few drawbacks. Some of the disadvantages are due to the early stage of the VLC technology and could be overcome in the future as the technology matures. The others are due to the usage of light as the medium and its fundamental characteristics. It will be difficult to completely mitigate the latter issues, but their effects could be alleviated and the communication techniques could be adapted to the situations. The main drawbacks of VLC are stringent LOS condition, limited transmission range, and susceptibility to interference from other light sources.

1.2 VLC in autonomous vehicles and ITS

The United Nations estimates that more than 1.3 million people die every year while 20 to 50 million are injured because of vehicle accidents. It is estimated that by 2020 road accidents will kill 1.9 million victims yearly [1]. In this context, reducing road accidents is a very serious concern. The United Nations declared in 2010 a Decade of Action for Road Safety.

In addition to vehicular safety systems, there is a strong need for environmental awareness of vehicles. This can be realized by using information obtained from different vehicles and infrastructure. This will not only increase the safety, but also to enhance the overall traveling experience. Such an intelligent transportation system (ITS) requires high cooperation among intelligent vehicles and between vehicles and intelligent transportation infrastructure, which will transform the whole transportation industry [2]. Needless to mention, ITS would combine advanced vehicle-to-infrastructure (V2I) and vehicle-to-vehicle (V2V) communication technologies for data gathering and distribution. Figure 1.1 shows the typical vehicular-VLC based scenario.

[2] Also, note there will be no multipath fading in an IM/DD environment because the power is always additive. However, there will still be multipath dispersion and inter symbol interference (ISI).

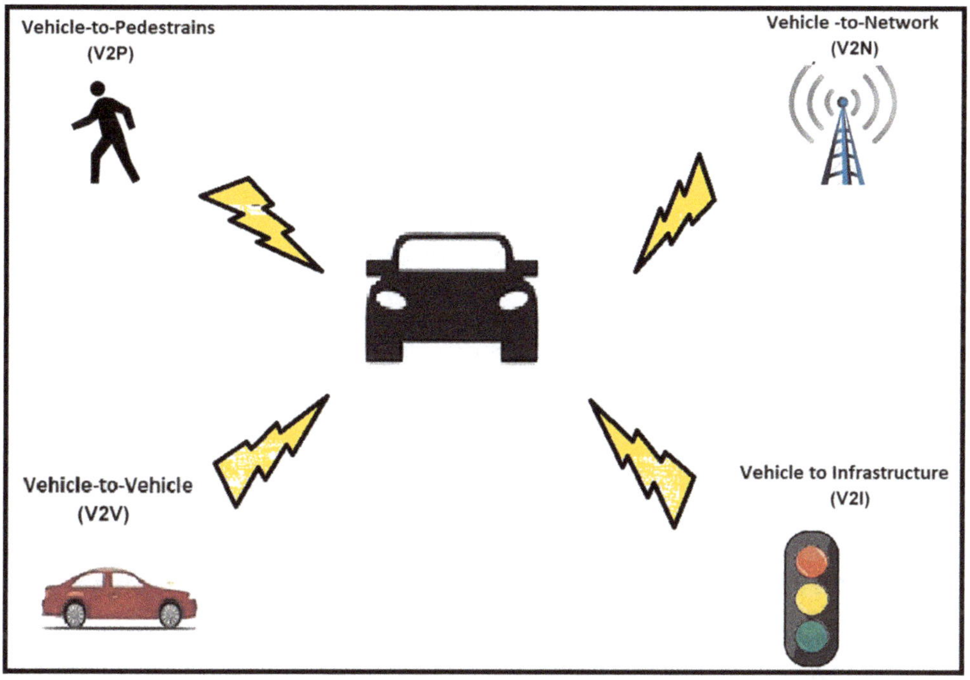

Figure 1.1. Application scenario for vehicular-VLC based.

Upcoming autonomous vehicles (AVs) will enable the ultimate realization of an ITS. Driverless vehicles are an incredibly disruptive technology, and as some people predict, autonomous driving is the third transportation revolution. The changes caused by widespread adoption of AVs stretch further than just transport. Entire urban environments have been constructed to suit cars as we know them. Cities, in future, may not need parking lots, taxi services or even public transport, because driverless fleets will replace them all. The savings in space and driving time will be huge. It will disruptively transform the way we live.

It has been suggested that one driverless car is likely to replace as many as 15–20 traditional cars. Couple this with electric vehicles (which, incidentally, is exactly what Tesla Motors is doing) and suddenly there is increased efficiency and less pollution due to lower fuel consumption. This will definitely create a very positive impact on air pollution in addition to space and time saving. However, there are still many technical hurdles to overcome before realizing the fully fledged benefits.

In particular, for both autonomous driving and ITS real time vehicular communication is imperative. In particular, in public traffic, where multiple other vehicles evolve in the traffic scene with changing velocity and positions. Due to traffic volume and limited lane resources, vehicles frequently interact with neighboring vehicles while moving along roads. This leads to highly dependent motion across vehicles. Communicating with them about our own immediate navigational intentions as well as understanding their intentions requires reliable, high-speed, wireless communication networks with very low latency.

Therefore, V2X (where, X stands for vehicles, infrastructure, networks, cellular base station, road side units (RSUs) such as traffic lights or ppedestrians) communications is essential for both the ITS and AVs. Realizing the importance, 5G wireless network standards provide specific features tailored towards AVs and ITS wireless communication such as *ultra reliable* and *low latency* modes.

The original V2X standard is based on IEEE 802.11p, a variant of the common wi-fi and part of the IEEE's WAVE, or Wireless Access for Vehicular Environments program, running in the unlicensed 5.9 GHz frequency band. This covers non line-of-sight-limited sensors such as cameras, radar and LIDAR, and covers V2V and V2I use cases such as collision warnings, speed limit alerts, and electronic parking and toll payments. Functional characteristics of 802.11p include short range (under 1 km), low latency (2 ms) and high reliability—basically, 802.11p extends a vehicle's ability to 'see' the environment around it, even in adverse weather.

An upcoming alternative to IEEE 802.11p is C-V2X, or cellular V2X. A key advantage of C-V2X is that it has two operational modes:

1. Low-latency C-V2X direct communications which is designed for active safety messages such as immediate road hazard warnings and other short-range V2X, V2I, and V2V situations, similar to IEEE 802.11p.
2. Communications on the regular licensed-band cellular network that can handle vehicle-to-cellular networks use cases like infotainment and latency-tolerant safety messages concerning longer-range road hazards or traffic conditions.

In this context, VLC can be a very productive complementary technology to IEEE 802.11p. VLC techniques can prove to be useful in specific scenarios, such as communicating with closely following vehicles and RSUs. In general, VLC and 802.11p are suitable technologies to support cooperative awareness; cooperative sensing and cooperative maneuver. On the other hand, the performance of IEEE 802.11p will be degraded in high load and dense network conditions for which VLC can be a handy supplement. VLC is also suitable for use cases such as see-through and platooning using short range links between a leading and following vehicle.

Due to the maturity in LEDs technology, car manufacturers have been replacing the classical halogen lamps with LED lighting systems. LEDs are highly reliable, energy efficient and have a lifetime that exceeds by far the classical light sources. Nowadays, and as illustrated in figure 1.2, vehicles' lighting systems based on LEDs are common.

Moreover, LED-based traffic lights have become more popular and are gaining usage on an extended scale. Normally a large number (100–200) of LEDs are used for a traffic light that offer the possibility for data transmission in addition to signaling function. Also, most of the street lighting will be LED-based in the near future, enabling potential infrastructure-to-vehicle (I2V)-VLC communication.

From the above, one can see that LED-based lighting will be part of the transportation system, being integrated with vehicles and in the infrastructure. The large geographical area in which LED lighting will be used, combined with VLC technology will allow ITS to gather data from a widespread area and can

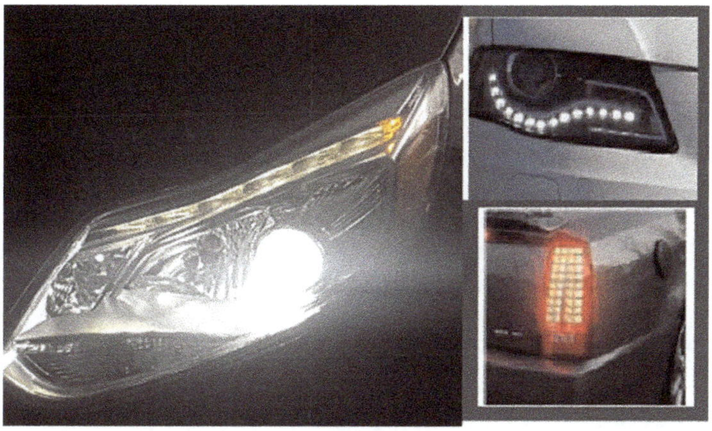

Figure 1.2. Integration of LED lighting systems in series vehicles.

enable the distribution of high-quality communications. These additional functions will be possible without affecting the primary goal of signaling or lighting in any way. This leads to the fact that VLC is able to satisfy the requirements imposed in vehicular networks in real working conditions as confirmed.

1.3 Research in the usage of VLC in AV and ITS

Since 1995, academic and industrial societies have been studying V2X and communication systems. Recently, VLC has been considered as a possible solution to enable V2X. The main advantages of VLC usage in automotive applications are represented by low complexity, reduced implementation cost and ubiquitous character. These characteristics can facilitate rapid market penetration, which represents a strong consideration in favor of VLC.

Many research groups have investigated the usage of VLC in automotive communication to increase traffic safety by allowing V2X communications. The simulation and experimental results obtained so far, prove that the use of VLC for road safety applications is possible, but to enable widespread implementation, the performances of such systems still need more improvement [3].

In the following paragraphs, we present a brief review of the work of some of the most representative research groups in the area of VLC automotive usage.

The group of Smart Lighting Engineering Research Center of Boston University (USA) is involved in the usage of VLC for inter-vehicular communication. The research team has developed a prototype used for vehicular networking based on optical transceivers. The system uses short-range directional optical transceivers to share vehicle state data. Also, they performed a comparative analysis between omnidirectional 802.11 RF communications and directional VLC, with application in vehicular communications [4]. The results show that in high traffic density, VLC offers better performance in terms of packet delivery ratio, throughput, and average packet delay, at the cost of a shorter communication range [5].

The group of professor Knightly from Rice University (USA) is another research group in VLC based vehicular communication. They developed a research platform with high robustness to noise on which they made experiments under different conditions in different scenarios. The experimental results show that VLC offers the possibility of robust vehicular communication in real traffic conditions. Experiments showed that a V2V–VLC system is resilient to sunlight noise except for the case of a direct line of sight with the Sun. Also, their results show that in dense traffic conditions, VLC satisfies the latency and the availability requirements imposed by the vehicle safety applications [6].

G Pang from the university of Hong Kong (China) and his research group are one of the pioneers of VLC systems for traffic safety. He showed that the LED has the potential to replace the classical light source and to be integrated with traffic lights, traffic signaling devices or into traffic display boards due to the LED's fast switching ability. This group proposed the replacement of all traffic lights and signaling devices with LED to reduce power consumption and to increase traffic safety. Also, they confirmed that the LED's long life expectancy, even in unfriendly working conditions, is extremely important since a burned traffic light can be a major risk factor. On the other hand, the communication capabilities of the LED can be used to further increase the safety of the transportation. Under these circumstances, the authors demonstrated the dual use of LED in the ITS: signaling and communication. To support their arguments, the research group presented a prototype traffic light whose purpose also transmits audio information through visible light [7]. Moreover, the research group proposed an intelligent traffic light system for the broadcasting of vehicle location and navigation information with audio support for the driver. The same group presented a different approach for traffic light to vehicle communication in [8].

The researchers from Nagoya University (Japan) are one of the early groups working in the VLC–ITS area. They have an inspiring background in the development of ITS-VLC base. In 2005 they proposed a parallel VLC system meant to broadcast traffic safety information from an LED-based traffic light to a high-speed camera-based receiver [9]. In 2007, they proposed a novel concept of VLC receiver prototype that aims to solve the main problems associated with the use of VLC in the ITS such as the necessity of long-distance high-speed transmission under the dynamic conditions [10]. The experimental results showed 2Mb/s bit rate at bit error rate (BER) below 10^{-6} for distances up to 40 m. The performance of the high-speed camera-based receivers and their decoding abilities have been improved in the years that followed. Later the communication distance was increased up to 120 m at BER of 10^{-2} in [11].

The collaboration between Shizuoka University (Japan) and the Applied Optics laboratory from Toyota enabled the development of a high-performance VLC system. They have developed a high sensitivity complementary metal-oxide semiconductor (CMOS) image sensor which is able to achieve 1000 (frame/second) rate. With the integration of the image sensor in a VLC receiver, their V2V prototype was able to achieve data rates of 10 Mb/s for distances up to 20 m [12]. The communication range of the systems could be increased up to 50 m by decreasing the data rate to 32 kb/s, or even up to 100 m for data rates of 2 kb/s [13].

One of the leading groups in the research of VLC usage for automotive applications is in Portugal, at the university of Aveiro. The group has proposed and analyzed in detail the use of LED-based traffic lights as RSU as a part of ITS for I2V data broadcast. The research group has considered the usage of photo detector based receivers and proposed the use of discrete sequence spread modulation (DSSS) inverse keying as a modulation technique [14]. The experimental results showed that this modulation technique is suitable for outdoor VLC since it reduces the effect of ambient noise [15]. The proposed system has been tested outdoors both at day and night times. The system was able to transmit 20 kb/s data up to 10 m at a BER of 10^{-6} and 45 m at 10^{-2}.

Although the discussed methods enhance the communication data rates, to date very high data rates (more than a Gbit/s) are not achievable even with advanced signal processing techniques. Also, providing a high-speed uplink is an additional challenge for VLC [16]. Another challenge for VLC is to cooperate with other RF wireless standards. The work in [17] examined the combination of a high-speed down link and a lower speed RF wireless LAN and showed that the combined system possesses some benefits in terms of latency and throughput. Also, the work in OMEGA project aims to combine different RF and optical wireless to achieve the desired performance [18].

1.4 Standardization efforts

1.4.1 IEEE 802.15.7 VLC: modulation schemes and dimming support

A number of standards have been developed and many are in the process for OWC. In [19], the IEEE 802.15.7 group showed that the integration of lighting and communication provides significant potential for VLC technology. They reported two main challenges in communication in this scenario, flicker mitigation and support for dimming. Moreover, they presented a few techniques to mitigate flicker and support dimming as defined in the IEEE 802.15.7 VLC standard. Besides, they found that there are many technical challenges that must be addressed to realize the full potential of VLC technology. The first challenge is that the channel models for VLC are not well understood, especially for NLOS for outdoor environments, and there is an active area of research for channel models and platforms for VLC.

The networking of the light sources and upgrading current infrastructures to support communication is another challenge, which requires support from the lighting industry. With continued growth in LED-based light sources and the need for multi-Gb/s data distribution, VLC, when developed as a global industry standard by IEEE 802.15.7, promises to be a very attractive candidate as a future high data rate and power-efficient technology.

Furthermore, using VLC in outdoor applications still faces many problems since the channel is highly affected by the meteorological conditions, besides the high noise from the Sun and other lighting sources. Therefore, the VLC system cannot be used as a long distance communication system.

1.4.2 Li-Fi

Li-Fi is defined as an optical wireless broadband access technology that uses the visible and infrared light spectrum to provide bidirectional capability. It supports simultaneous lighting and communications in the VLC band and also point-to-point or point-to-multipoint topology in both uplink and downlink directions. Li-Fi also supports multiuser access and offers mobility support for both intra-cell and inter-cell movement using traditional (horizontal) handover. Li-Fi has an added feature (vertical handover) that enables the ability to maintain a data connection to the network as a user changes the principle radio access technology, i.e., moving from 4G to wi-fi or from wi-fi to Li-Fi.

The Li-Fi infrastructure comprises multiple LEDs that form a wireless communication network, offering a typical MIMO scenario. The developers of Li-Fi systems envision interoperability with wi-fi, 4G or other RF systems where a user can be 'opportunistically', and even simultaneously connected to Li-Fi and other RF wireless systems.

1.4.3 IrDA

The Infrared Data Association (IrDA) communication standards have been around since 1993. IrDA provides specifications for wireless infrared communications in the 850–900 nm band. IrDA has been implemented in portable devices such as mobile telephones, laptops, cameras, printers, and medical devices. There are several flavours of IrDA offering from 2.4 kb/s to 1 Gbit/s (GigaIR) covering from a few centimeters to a few meters. IrDA typically works in half-duplex mode (or unidirectional). The trade-off with this technology is between distance and bit-rate.

1.4.4 IEEE 802.11bb™

In May 2018, IEEE announced the formation of the IEEE 802.11™ Light Communications Task Group to develop a global standard for light communications in wireless local area networking. The IEEE 802.11bb™ task group is to define one medium access control (MAC) and several physical layer (PHY) specifications for light-based wireless connectivity for fixed, portable, and moving stations within a local area network. The new standard will closely align with Li-Fi.

The key specification of IEEE 802.11bb™ is: uplink operation is in the 380 nm band while downlink operation is in the 5000 nm band. Also, all modes of operation of IEEE 802.11bb™ are supposed to achieve minimum single-link throughput of 10 Mb/s. Meanwhile, at least one mode of operation should achieve single-link throughput of at least 5 Gb/s. Interoperability among solid state light sources with different modulation bandwidths should also be supported.

1.4.5 Safety concerns of OWC systems

Infrared, visible or ultraviolet electromagnetic radiation, in sufficient quantity or concentration, can cause damage to the human eye. Many cases of laser induced eye damage have been documented. The visible light from LED is usually safe as an unknown exposure rarely happens. On the other hand, IR cannot be seen and it can damage the retina of the human eye. The amount of power that can be emitted by an IR radiator is controlled by several regulating bodies . At high intensities it may even damage our skin. IEC 825-1 (Compliance of Infrared Communication Products) is the dominant standard that regulates the MPE (maximum permissible exposure)[3], as a function of source area, exposure duration and wavelength. Typically, MPE is higher at longer wavelengths with bigger source area.

The initial standard, IEC 825 was first published in 1984 and was based on the American National Standard ANSI Z-136.1. The most significant change that took place with the second edition of the standard (i.e., IEC 825-1) has been to include LEDs since recently the power emission capabilities of LED have significantly gone up.

1.5 The architecture of the VLC system

A VLC system consists of a transmitter to generate modulated light and a receiver to detect the received light variation. Both the transmitter and the receiver are connected through the wireless channel. Figure 1.3 illustrates the basic architecture of a VLC system.

1.5.1 VLC emitter

A digital VLC emitter is a device that converts the binary data to intensity modulated light waves for transmission. A driving circuit controls the switching of the LED according to the incoming binary data at the given data rate, generating an amplitude modulated light beam. In the simplest case, the light produced by the LED is modulated with ON–OFF-keying (OOK) amplitude modulation. Since the amplitude can be easily corrupted by additive noise, amplitude-independent techniques such as pulse width modulation (PWM) or pulse position modulation

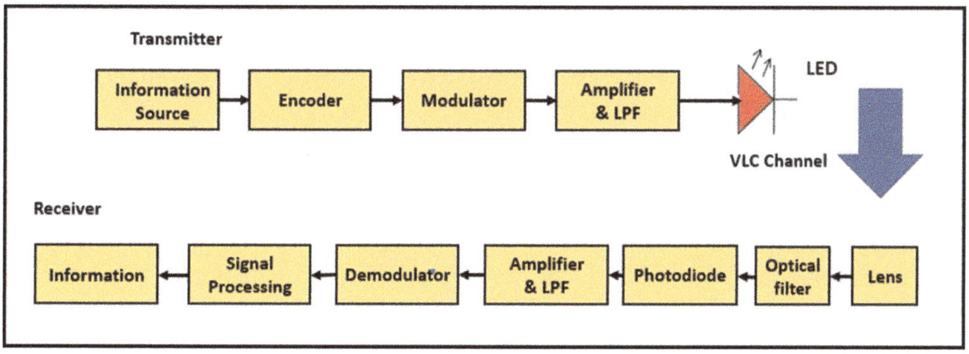

Figure 1.3. Block diagram of a VLC communication link.

[3] MPE is approximately 1/10th of the level of electromagnetic radiation exposure which will cause damage in 50% of human eyes.

(PPM) are often used in digital OWC systems. Advanced modulation techniques such as OFDM [20], discrete multi-tone modulation (DMT) or DSSS [14] can also be used that provide more advantages at higher complexity.

The VLC emitter has a dual purpose, emits light, and transmits data instantaneously by using the same optical power without any noticeable flickering, although the human eye cannot detect flicker as long as the modulation frequency is kept above the fusion threshold. However, the perceived intensity is changed depending on the relative periods of light and darkness (dimming effect). The flicker fusion threshold is a statistical rather than an absolute quantity. The flicker fusion threshold at very high illumination intensities is about 60 Hz.[4] However, this could be low at low intensities. While we may not see the flicker, the dimming effect is real. Note that the LED brightness, while it is on has to be twice as high if a 50% on–off rate is used.

VLC uses typically white light LEDs common for lighting. White light generation using solid state technology did not mature until recently since a PN junction emits only one wavelength. Currently, two techniques are used for white light generation: R–G–B LED, and phosphorescent LED. The R–G–B LED consists of red, green, and blue chips and combines the three lights in correct proportion to generate white light. R–G–B LED has relative higher modulation bandwidth, up to 20 MHz and can support wave division multiplexing (WDM), but the cost is relatively high. Phosphorescent LED uses the blue LED chip coated with a yellow phosphor, which is the most popular white LED in the market due to its low cost. Phosphor type LEDs have turn-on and turn-off times in the tens to hundreds of nanoseconds, appreciably slower than direct emission LEDs [21].

In general, the data rate of the VLC link depends on the switching capabilities of the LEDs while the emitter's area controls the transmission power. Direct emission LEDs typically have a turn-on time in single-digit nanoseconds, longer for bigger LEDs. However, turn-off times for these are in the tens of nanoseconds, slower than turn-on times. This is because, typical LED driving topologies do not actively pull the voltage across the LED down when turning off. Special purpose LEDs are available at a high cost, whose junction and bond-wire geometries are designed specifically to permit few hundred picosecond pulses.

In practice the limiting time constants can be in the hundreds of nanoseconds. This is largely due to external factors such as:

- Inductance of the traces. The longer the traces, the slower the transitions due to inductance.
- Junction capacitance of the LED.
- Parasitic capacitance (traces and support circuitry) plays an important role in increasing the RC time constant and thus slowing transitions.

As a result of the inductive and capacitive factors above, the higher the forward voltage of the LED, the longer the rise and fall times, due to need for higher driving currents.

[4] This is why we don't see the flicker of florescent light.

However, the typical bit rates that can be supported by fast moving vehicles is usually limited by channel conditions and ISI, not by the switching speed of the LED.

1.5.2 VLC receiver

The VLC receiver will be detecting the data from the received modulated light beam and transforms it to an electrical signal that will be decoded. Usually, the PD has more modulation bandwidth than LED, and the signal-to-noise-ratio (SNR) of the receiver is the primary design constraint.

There are two widely adopted types of PDs, ordinary positive intrinsic negative photodiodes (PIN) and avalanche photodiodes (APD). Major signal and noise elements of a generalized PD are shown in figure 1.4. Here G is the avalanche gain and $F(G)$ is the excess noise of the APD. For PIN, we may assume $G = F(G) = 1$ and use the same diagram without losing the generality. The dark current noise, $\langle I_{\text{dark}}^2 \rangle = \sigma_{\text{Dark}}^2$ is usually negligible in VLC environments. Thermal noise $\langle I_{\text{th}}^2 \rangle = \sigma_{\text{Thermal}}^2$ arises from the bias resistance and a constance irrespective of the received signal. The shot noise, $\langle I_{\text{shot}}^2 \rangle = \sigma_{\text{Shot}}^2$ is proportional to the received optical power. Note, the received signal power includes both the energy transmitted from the OWC transmitter and from ambient light. Often the ambient light is of the same order or even larger than the transmitted signal because of the open nature of the channel. Hence, the shot noise can be large and it will often dictate the performance of an OWC receiver.

Careful observation of figure 1.4 reveals that the inner (avalanche) gain of the APD multiplies the received signal (and the shot noise) before the thermal noise is added[5]. Note the quantum noise is proportional to received signal plus noise power within the detection bandwidth. Hence, in contrast to fiber optic systems, OWC systems have large in-band ambient noise resulting in large shot noise. Therefore, the APD, although an expensive device with built-in gain, is not favored in OWC systems since there is large ambient-induced shot noise.

The key parameter in PDs is the responsivity γ which is defined as [22],

$$\gamma = \frac{I_p}{P_{\text{inc}}} \tag{1.1}$$

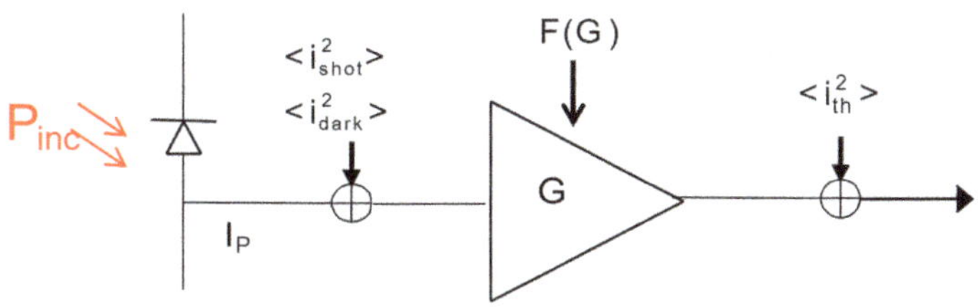

Figure 1.4. Signal and noise elements of a generalized photodiode.

[5] Quantum and shot noise are the same.

where I_p denotes the average generated photocurrent and P_{inc} denotes the incident optical power.

The PD responsivity depends on the physical structure of the PD and has the units of (A/W). The incident power P_{inc} is proportional to its effective light collection area.

Thus, the PD effective area must be large enough to collect enough transmitted signal power. However, a large area will increase the junction capacitance of the PD slowing it down. Another factor that impacts the speed is the electron transit time, the time taken for the electrons to travel across the depletion layer. The wider the depletion layer, the higher would be the transit time. However, a wider depletion layer would absorb more photons and will increase the responsivity. Therefore, designing an appropriate PD for OWC applications is quite different from designing it for fiber optic applications. Fiber optic detectors would have a very small photosensitive area and be expected to work at Gigabit speed, while the OWC detectors would have relatively large area, and still be expected to work at high speed.

Note that the incident light is not only due to the emitter but also from other light sources (artificial or natural), the receiver is subject to significant interferences. The performance of the VLC receiver can be enhanced using an optical filter that rejects the unwanted spectral components of the interfering noise signal. In general, the VLC receiver must be selected such that the cost, performance and sensitivity requirements are satisfied [22].

1.5.3 VLC channel

The transmitter and the receiver of the VLC system are connected through the free space optical communication channel. At the receiver, the received power intensity decreases with the square root of the distance between the transmitter and the receiver (assuming a free space propagation model). This makes the received signal usually low. Also, the VLC channel could contain many sources of optical noise. During the daytime, the most dominant source of the channel noise is the Sun. Different noise sources and the weather condition make the VLC channel unpredictable. Rain, snow, dust particles or heavy fog can affect the VLC link quality by causing scattering of the light containing the data. Noise sources, together with the low signal power especially at a long distance, significantly affects the SNR in VLC systems.

Another main characteristic of the VLC channel is directional propagation of light rays that result in the LOS channel to be dominant. The multipath (reflected) signal collected at the receiver has limited effect, which is experienced only at short emitter–receiver distances. The VLC channel is discussed in detail in the next chapter.

1.5.4 Comparison of RF and VLC propagation channels

Both radio and visible light waves can be used to enable communications in highly dynamic vehicular environments. Unfortunately, the roles of these two technologies and how they interact with each other in future vehicular communication systems remain unclear. Understanding the propagation characteristics is an essential step in investigating the benefits and shortcomings of each technology.

Comparison of relevant properties such as radiation pattern, path loss modeling, noise and interference, and channel time variation of radio and visible light propagation channels, provides an important insight that the two communication channels can complement each other's capabilities in terms of coverage and reliability, thus better satisfying the diverse requirements of future cooperative intelligent transportation systems. table 1.1 summarizes the most important observations for applications stemming from propagation channel perspective.

1.6 Noise and SNR

In OWC systems, the most important challenge is the background noise which is created from ambient light in the environment. Ambient light can rise from sunlight, skylight, advertising signs and fluorescent lamps, or other light sources [23]. The background noise existing in the environment leads to a DC photocurrent that causes a shot noise at the receiver. This noise source is signal independent and is the result of the summation of many independent random variables. Hence, it can be modeled as a Gaussian distribution [22].

The shot noise variance is given by [24]

$$\sigma_{\text{Shot}}^2 = 2q_e G^2 F(I_{\text{inc}})B \tag{1.2}$$

where F denotes the excess noise, q_e denotes the electronic charge constant, G is a PD gain, B denotes the PD bandwidth, and I_{inc} denotes the photocurrent produced due to the received optical power.

The photocurrent I_{inc} is given by [25]

$$I_{\text{inc}} = GA_r \int_{\lambda_1}^{\lambda_2} P_{\text{inc}}(\lambda)\gamma(\lambda)T_o(\lambda)d\lambda \tag{1.3}$$

Table 1.1. Comparison of RF and VLC propagation channels for vehicular communications.

Item	Description
Range	Radio channel is used for longer range and optical channel for shorter range applications
Datarate	Optical channel has the potential to deliver a high per-link bit rate and can even sustain long link durations
Complexity	Optical receiver signal-processing complexity of vehicular VLC is much lower than its RF counterpart
Coverage	VLC communication ensures coverage of only a small area. Therefore, it is less suitable for use cases which require omnidirectional transmission
Cost	Radio-based solutions have higher initial deployment costs. Vehicular VLC has a lower initial cost as LEDs are common in vehicles today and additional electronics to enable VLC is inexpensive

where λ_1, λ_2 are the visible light wavelengths, $\gamma(\lambda)$ is the responsivity of the APD given in A/W at certain λ, A_r is the effective area of the receiver, and $T_o(\lambda)$ is the transmittance of the bandpass optical blue filter.

In addition to the shot noise, there are many different types of noises that are liable to degrade the performance of the VLC system, such as dark noise and thermal noise. Dark noise is given by [24]

$$\sigma^2_{\text{Dark}} = 2q_e G^2 F I_{\text{dg}} B + 2q_e I_{\text{ds}} \tag{1.4}$$

where I_{ds} is a surface dark current, and I_{dg} is a bulk dark current.

Also, due to a random thermal motion of charge carriers, thermal noise is generated. The variance of thermal noise is given by [24]

$$\sigma^2_{\text{Thermal}} = 4\left(\frac{K_{\text{B}} T_{\text{k}}}{R_{\text{L}}}\right) F_{\text{n}} \text{B} \tag{1.5}$$

where K_{B} denotes Boltzmann's constant, T_{k} denotes absolute temperature, R_{L} denotes the load resistance, and F_{n} is the photodiode noise figure.

Assuming the noise processes are independent to each other, the variance of the ambient total noise σ^2_{n} is given by

$$\sigma^2_{\text{n}} = \sigma^2_{\text{Shot}} + \sigma^2_{\text{Dark}} + \sigma^2_{\text{Thermal}} \tag{1.6}$$

table 1.2 shows the probability density function (PDF) of each type of noise.

In this book, Hamamatsu-S8664 APD is considered as a receiver. The simulation parameters and their values are given in table 1.3 [29].

1.7 Chapter summary

This chapter introduces the concept, the architecture design, the advantages and the drawbacks of VLC system. Also, different types of noise and SNR are introduced. The potential usage and the role of the VLC in an ITS have been discussed. In an ITS, VLC appears to be the solution, especially for urban high-traffic densities. This chapter also highlights the current trends in the development of VLC systems and the challenges in the domain.

Table 1.2. PDFs of different types of noise.

Noise	PDF
Shot noise	Poisson distribution [26]
Dark noise	Poisson distribution [26]
Thermal noise	Gaussian distribution [26]
ISI noise	Gaussian distribution [27]
Clipping noise	Gaussian distribution [28]

Table 1.3. Noise parameters and their simulation values.

Symbol	Description	Value
APD model	(Si) Hamamatsu	*S* 8664–1010
λ	Visible light wavelength	350–750 nm
G	APD gain	50
q_e	Electron charge	1.9×10^{-19} C
B	APD bandwidth	65 MHz
I_{ds}	Surface dark current	10 nA
I_{dg}	Bulk dark current	5100 μA
K_B	Boltzmann's constant	1.38×10^{-23} J/K
T_k	Temperature	298 K
F_n	APD noise figure	0.2
γ	Photodetector responsivity	0.28 A/W
R_L	Load resistance	50 Ω

References

[1] World Health Organaization www.who.int/news-room/fact-sheets/detail

[2] Papadimitratos P, La Fortelle A D, Evenssen K, Brignolo R and Cosenza S 2009 Vehicular communication systems: Enabling technologies, applications, and future outlook on intelligent transportation *IEEE Commun. Mag.* **47** 84–95

[3] Căilean A-M 2014 Study, implementation and optimization of a visible light communications system: application to automotive field *PhD Thesis* l'Université de Versailles Saint-Quentin-en-Yvelines

[4] Agarwal A and Little T D C 2010 Role of directional wireless communication in vehicular networks *IEEE Intelligent Vehicles Symp. (June 2010)* pp 688–93

[5] Little T D C, Agarwal A, Chau J, Figueroa M, Ganick A, Lobo J, Rich T and Schimitsch P 2010 Directional communication system for short-range vehicular communications *IEEE Vehicular Networking Conf. (Dec 2010)* pp 231–8

[6] Knightly B, Liu E W and Sadeghi C 2011 Enabling vehicular visible light communication (V2LC) networks *Proc. of the Eighth ACM Int. Workshop on Vehicular Inter-networking (VANET'11)* (New York: ACM), pp 41–50

[7] Chan C H, Liu H, Kwan T and Pang G K H 1998 Dual use of LEDs: signaling and communications in ITS *Proc. of fifth World Congress on Intelligent Transport Systems, Paper 3035 (Seoul, Korea)* pp 12–6

[8] Liu H S and Pang G 2003 Positioning beacon system using digital camera and LEDs *IEEE Trans. Veh. Technol.* **52** 406–19

[9] Wada M, Yendo T, Fujii T and Tanimoto M 2005 Road-to-vehicle communication using LED traffic light *IEEE Proc. Intelligent Vehicles Symp. (June 2005)* pp 601–6

[10] Hara T, Iwasaki S, Yendo T, Fujii T and Tanimoto M 2007 A new receiving system of visible light communication for ITS *IEEE Intelligent Vehicles Symp.* pp 474–9

[11] Nagura T, Yamazato T, Katayama M, Yendo T, Fujii T and Okada H 2010 Improved decoding methods of visible light communication system for ITS using LED array and high-speed camera *IEEE 71st Vehicular Technology Conf. (May 2010)* pp 1–5

[12] Takai I, Ito S, Yasutomi K, Kagawa K, Andoh M and Kawahito S 2013 LED and CMOS image sensor based optical wireless communication system for automotive applications *IEEE Photonics J.* **5** 6801418

[13] Yamazato T *et al* 2014 Image-sensor-based visible light communication for automotive applications *IEEE Commun. Mag.* **52** 88–97

[14] Terra D, Kumar N, Lourenço N, Alves L N and Aguiar R L 2011 Design, development and performance analysis of DSSS-based transceiver for VLC *IEEE EUROCON—Int. Conf. on Computer as a Tool (April 2011)* pp 1–4

[15] Kumar N, Lourenço N, Terra D, Alves L N and Aguiar R L 2012 Visible light communications in intelligent transportation systems *IEEE Intelligent Vehicles Symp.* pp 748–53

[16] O'Brien D C, Zeng L, Le-Minh H, Faulkner G, Walewski J W and Randel S 2008 Visible light communications: Challenges and possibilities *IEEE 19th Int. Symp. on Personal, Indoor and Mobile Radio Communications (Sept 2008)* pp 1–5

[17] Hou J and O'Brien D C 2006 Vertical handover-decision-making algorithm using fuzzy logic for the integrated Radio-and-OW system *IEEE Trans. Wirel. Commun.*

[18] OMEGA. Omega project www.ict-omega.eu

[19] Rajagopal S, Roberts R D and Lim S K 2012 IEEE 802.15.7 visible light communication: modulation schemes and dimming support *IEEE Commun. Mag.* **50** 72–82

[20] Elgala H, Mesleh R and Haas H 2011 On the performance of different OFDM based optical wireless communication systems *IEEE/OSA J. Opt. Commun. Net.* **3** 620–8

[21] Le Minh H, O'Brien D, Faulkner G, Zeng L, Lee K, Jung D, Oh Y and Won E T 2009 100-Mb/s NRZ Visible Light Communications Using a Postequalized White LED *IEEE Photonics Technol. Lett.* **21** 1063–5

[22] Hranilovic S 2005 *Wireless Optical Communication Systems* (New York: Springer)

[23] Yu S-H, Shih O, Tsai H-M and Roberts R 2013 Smart automotive lighting for vehicle safety *IEEE Commun. Mag.* **12** 50–9

[24] Komine T and Nakagawa M 2004 Fundamental analysis for visible-light communication system using LED lights *IEEE Trans. Consum. Electron.* **50** 100–7

[25] Lee S J, Kwon J K, Jung S-Y and Kwon Y-H 2012 Evaluation of visible light communication channel delay profiles for automotive applications *EURASIP J. Wirel. Commun. Net.* 1823–6

[26] Agrawal G P 2014 *Fiber-Optic Communication Systems* 4th edn (New York: Wiley)

[27] vanRooyen P and Solms F 1995 Maximum entropy investigation of the inter user interference distribution in a DS/SSMA system *Proc. of 6th Int. Symp. on Personal, Indoor and Mobile Radio Communications (Sept 1995)* **vol 3**

[28] Lu Q f, Ji X s and Huang K z 2014 Clipping distortion analysis and optimal power allocation for ACO-OFDM based visible light communication *4th IEEE Int. Conf. on Information Science and Technology (April 2014)* pp 320–23

[29] Photo diode parameters www.hamamatsu.com

Chapter 2

Channel modeling for the V2V–VLC system

Accurate VLC channel models are necessary for better system design. Using VLC in outdoor applications has a few challenges such as signal loss during severe weather conditions, sunlight, and ambient light. Weather conditions such as heavy fog, rain or snow could decrease the communication range. These degrading effects can be minimized by using highly sensitive receivers. Another strong concern is direct sunlight or strong ambient light, which could saturate the receiver [1]. Characterization of a communication channel is performed by its channel impulse response, which is then used to study, analyze, and combat the effects of channel distortions.

In literature, many works have been published on the channel characterization, covering both experimental measurement and computer modelling of indoor and outdoor OWC systems. Most of the optical wireless works in the VLC environment have focused on indoor channel modeling. There is little work on outdoor applications, especially in multipath mobile V2V channels. In outdoors, the power penalties directly associated with the channel may be separated into two factors, these being large optical path loss and multipath dispersion.

In this book, two types of configurations are considered in an optical wireless channel, LOS and NLOS. For directed LOS and tracked configurations, reflections do not need to be taken into consideration, and consequently the path loss is easily calculated from knowledge of the transmitter beam divergence, receiver size and separation distance. However, a NLOS configuration, also known as diffuse systems uses reflections off the surrounding objects surfaces. Reflections could be seen as unwanted signals or multipath distortions which make the prediction of the path loss more complex. A number of propagation models (ceiling bounce, Hayasaka–Ito and spherical) for LOS and NLOS are introduced in the literature [2], and will be given in brief in this chapter.

The atmospheric outdoor channel is a very complex and dynamic environment that can affect characteristics of the propagating optical beam, thus resulting in

doi:10.1088/978-0-7503-2284-3ch2

optical losses and turbulence induced amplitude and phase fluctuation. There are a number of models to characterize the statistical nature of the atmospheric channel presented in literature [2].

In this chapter, we introduce a novel V2V–VLC system in a rich scattering environment considering both LOS and NLOS paths. We employ a pair of headlights as transmitters in one vehicle, and a pair of taillights as receivers in the adjacent vehicle, making the system analogous to a 2 × 2 MIMO communication link. Also, we present a closed-form expression of the CIR, and various channel parameters are obtained.

2.1 Channel modeling for VLC

There are a number of topologies that are generally used for indoor applications. The configurations can be classified according to many factors such as the transmitted angle, the incident angle, and the LOS path between the transmitter and the receiver. The most commonly used technique for the optical communication is the intensity modulation with direct detection (IM/DD). Here, the driving current of an optical source is directly modulated by the modulating signal which in turn varies the intensity of the optical source, as shown in figure 1.3.

The receiver employs a PD, with a response which is the integration of tens of thousands of very short wavelengths of the incident optical signal that generates a photocurrent. This photocurrent is directly proportional to the instantaneous optical power incident on it, that is, proportional to the square of received electric field. An IM/DD-based optical wireless system has an equivalent baseband model that hides the high- frequency nature of the optical carrier.

The IM/DD model is shown in figure 2.1. The CIR $h(t)$ is a baseband CIR and $n(t)$ is the signal-independent shot noise, modelled throughout the book as the additive white Gaussian noise (AWGN) with a double-sided power spectral density (PSD) of $N_0/2$.

NLOS are subject to the effects of multipath propagation in the same way as RF systems and these effects are more noticeable. This type of link can suffer from severe multipath-induced performance penalties. Multipath propagation causes the electric field to suffer from severe amplitude fades on the scale of a wavelength. If the detector size is in the order of one wavelength, then the detector will experience destructive interference and fading.

In an IM/DD optical wireless, the received signal can be written as

$$y(t) = \gamma x(t) \otimes h(t) + n(t) \tag{2.1}$$

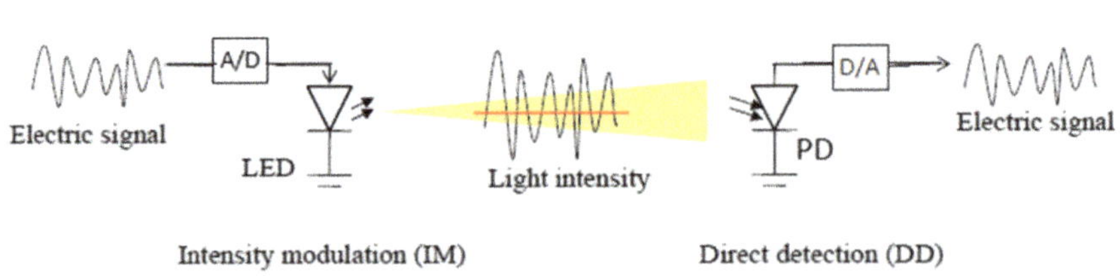

Figure 2.1. An optical intensity modulation, direct detection communications channel.

where $x(t)$ is the transmitted signal, $y(t)$ is the received signal, $h(t)$ is the CIR, γ is the PD responsivity, $\otimes$ is the convolution sign, and $n(t)$ is the noise.

While (2.1) is simply a linear filter channel with AWGN, optical wireless systems differ from conventional electrical or radio systems since the instantaneous optical power is proportional to the generated electrical current $x(t)$ represents the power rather than the amplitude signal. This imposes two constraints on the transmitted signal. Firstly, $x(t)$ must be non-negative, that is $x(t) \geqslant 0$. Secondly, the eye safety requirements limit the maximum optical transmit power that may be used. Generally, it is the average power requirement which is the most restrictive and hence, the average value of $x(t)$ must not exceed a specified maximum power value $P_{\max}$. This is different to the time-averaged value of the signal $|x(t)|^2$, which is the case for the conventional RF channels when $x(t)$ represents amplitude [2]. These differences have a reflective effect on the system design. On conventional RF channels, the SNR is proportional to the average received power, whereas on optical wireless links, it is proportional to the square of the average received optical signal power.

2.1.1 Indoor optical wireless communication channel models

There are a number of topologies that are commonly used for indoor applications. The configurations can be classified according to many parameters such as the degree of directionality of transmitter and receiver and the existence of the LOS path between the transmitter and the receiver. In this chapter, we will give a brief description of the indoor OWC channel models.

2.1.1.1 LOS propagation model

In this model, the DC gain for a receiver located at a distance of d and angle ϕ with respect to transmitter can be approximated as [3]

$$h_{\mathrm{LOS}}(0) = \begin{cases} \dfrac{A_r(m_l + 1)}{2\pi d^2}\cos(\phi)T_s(\Psi)g(\Psi)\cos(\Psi) & 0 \leqslant \Psi \leqslant \Psi_c \\ 0 & \text{elsewhere} \end{cases} \tag{2.2}$$

where m_l is the Lambert's mode number expressing infectivity of the source beam, ϕ is the angle of the radiated power, d is the distance between the transmitter and receiver, $T_s(\Psi)$ is the optical transmission coefficient, $g(\Psi)$ is the optical gain of an ideal non imaging concentrator, A_r is the receiver effective area, Ψ_c is the critical angle and Ψ is the FOV angle as shown in figure 2.2.

In short-distance LOS links, multipath dispersion is seldom a problem and LOS links channel are often modelled as a linear attenuation and delay [4]. The optical LOS links are considered as nonfrequency selective and the path loss depends on the square of distance between the transmitter and the receiver. The impulse response can be expressed as [3]

$$h_{\mathrm{LOS}}(t) = \frac{A_r(m_l + 1)}{2\pi d^2}\cos(\phi)T_s(\Psi)g(\Psi)\cos(\Psi)\delta\left(t - \frac{d}{c}\right) \tag{2.3}$$

where c is the speed of the light in free space and $\delta(.)$ is the Dirac function.

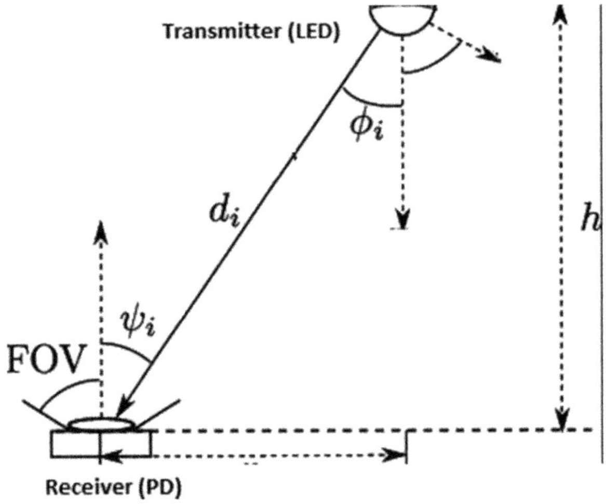

Figure 2.2. Geometry of the LOS propagation model.

2.1.1.2 NLOS propagation model

For NLOS and diffuse links, the optical path loss is more complex to predict since it is dependent on many of factors, such as room dimensions, the reflectivity of the ceiling, walls and objects within the room, and the position and orientation of the transmitter and receiver, window size and place and other physical matters within a room [2]. The received power consists of two components. the LOS power and the NLOS power, and is generally defined as

$$P_r = [H_{\text{LOS}}(0) + H_{\text{NLOS}}(0)]P_t \tag{2.4}$$

where P_r and P_t represent the received power and transmitted power respectively.

Mathematically, the impulse response of the optical wireless channel is calculated by integrating the power of all the components arriving at the receiver after multipath propagation [5]. The received signal in the case of the non-LOS links consists of various components arriving from different path, the path lengths of these components differ in proportion to the room design, hence there is broadening of the pulse. The distribution of the channel gain in decibels (dB) for the LOS component follows a modified gamma distribution, and the channel gain in dB for LOS channels including all reflections follows a modified Rayleigh distribution for most transmitter–receiver distances [6], where the CIR is given by (2.23). The root mean square delay spread is a parameter which is commonly used to quantify the time-dispersive properties of multipath channels, and is defined as the square root of the second central moment of the magnitude squared of the channel impulse response [4], which can be given later in this chapter by (2.50) and (2.51), respectively.

2.1.1.3 Ceiling bounce model

As mentioned before, many models had been proposed and studied for simulating impulse response of indoor IR channel. Carruthers and Kahn presented the ceiling bounce model. This model is commonly used in simulation because of its excellent matching with the measured data and simplicity of the model. This method came up with a closed-form expression for the impulse response assuming the transmitter and the receiver to be collocated in planes parallel to the floor and directed towards the ceiling. In the ceiling bounce model, the multipath IR channel is characterized by only two parameters: the optical path loss and the RMS delay spread. The impulse response $h(t)$ of the ceiling bounce model is given by [2]

$$h(t, a) = H(0)\frac{6a^6}{(t + a)^7}u(t) \tag{2.5}$$

where $u(t)$ is the unit step function and a is the height of the ceiling above the transmitter and receiver.

2.1.1.4 Hayasaka–Ito model

In this model, the diffuse IR channel is analysed by the ray-tracing technique and the modified Monte Carlo method. Assuming a Lambertian model for the light sources, the impulse response of the channel is expressed as [7]

$$h(t) = h_1(t - T_1) + h_{\text{high}}(t - T_{\text{hight}}) \tag{2.6}$$

where $h_1(t)$ is the impulse response of the primary reflection, $h_{\text{high}}(t)$ is that of the higher order reflections, T_1 and T_{high} are the starting times of the rise of the primary and higher order reflections [2].

2.1.1.5 Spherical model

The spherical model uses with higher order reflections [8] with a gain given by

$$h_{\text{high}} = \frac{\eta_{\text{high}}}{\tau}e^{\frac{-t}{\tau_e}} \tag{2.7}$$

where the power efficiency of the diffuse signal η_{high}, the exponential decay time τ_e and the average transmission delay $<t>$ for one reflection are given by [2].

2.2 Channel modeling for V2V–VLC system

The development of vehicular communications requires accurate channel models so as to evaluate the system's performance. Channel modeling in a V2V environment is quite complex due to high relative velocities between the transmitter and the receiver in conjunction with a dynamic variation of nodes' position. Another significant characteristic is the high Doppler spread/shift that results in a statistically non-stationary channel.

2.2.1 History of channel modeling of vehicular communication

A mathematical model for the V2V–VLC system was presented in [9]. They used a market-weighted headlamp beam pattern considering both the LOS and the NLOS links. They considered the impact of NLOS path coming from the road surface reflection only. Also, they considered good weather conditions in their analysis. In [10], the authors extended their work in [9], but they assumed a wet surface road and the reflection is not Lambertian. They investigated the system's performance in terms of BER and the hight of the receiver.

The authors in [11] proposed a measurement-based time variant non-clear sky channel model for the I2V–VLC system. Also, they consider the dynamic characteristics of background radiation to enable more realistic and accurate prediction of the VLC system performance in an outdoor application. Moreover, they introduced a new receiver design with dual-reception and effective ambient-light rejection capabilities which employs the selection diversity technique in order to mitigate the impact of ambient-light noise due to daylight.

In [12], a time function of the V2V channel path loss caused by vehicle mobility had been obtained using the video data collected from cameras mounted on driven vehicles. Computer vision techniques were used to identify the location of the taillights of the transmitting vehicle with respect to the receiving vehicle.

The authors in [13] used a ray-tracing scheme employing commercial light tools software to evaluate channel delay profiles obtained from CIR for multiple LOS links and fewer NLOS delay taps for crossroad and metropolitan street under the V2V and V2I communication links. They concluded that the delay profile from the V2I link and metropolitan scenario had more dispersive channel characteristics due to the reflection and the diffusion of the visible light.

In [14], a channel model of OWC systems during rainfall was proposed. The authors considered outdoor optical wireless systems where optical transmitter communicates directly along point-to-point LOS propagation links. They considered the scattering of optical light through raindrops in the propagation channel and derived the PDF of the received signal power.

The authors in [15] evaluated the VLC channel model in the crossroad and metropolitan scenarios based on the practical VLC light sources. They used light tools and CATIA V5 tools to simulate their work.

In [16], the authors used an image sensor instead of PDs as a receiver, where this technique is considered as an attractive solution for the outdoor mobile applications. They introduced the pinhole camera model concept to express motion of the VLC transmitters. Moreover, they used the motion models to simulate and analyze VLC channel fluctuation.

2.3 (2 × 2) MIMO V2V–VLC system model

This section discusses the system model of V2V–VLC in detail. The V2V–VLC system which is shown in figure 2.3 is comprised of two vehicles communicating with each other by utilizing headlights and taillights. Headlights of the first vehicle act as a transmitter and taillights of the second vehicle act as a receiver. Since each vehicle

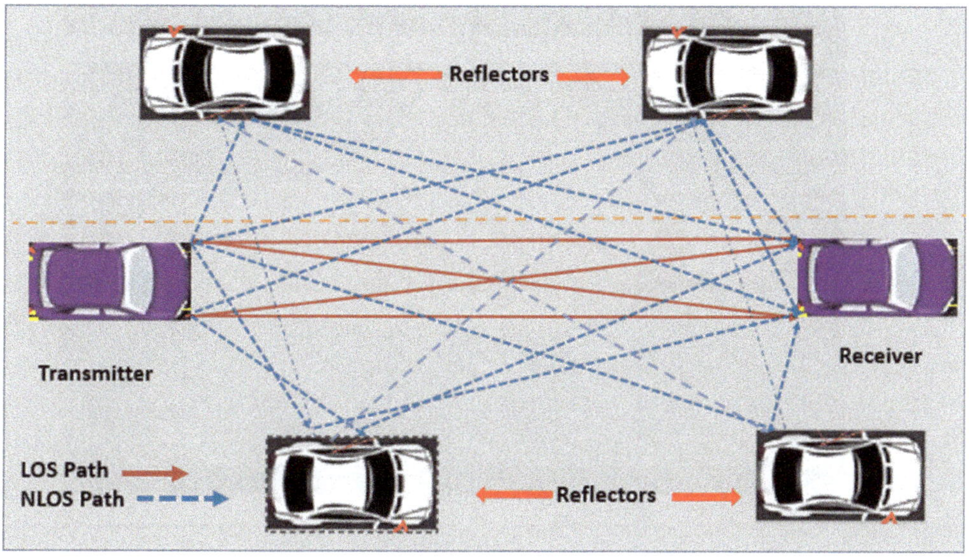

Figure 2.3. (2 × 2) MIMO model for V2V–VLC showing the LOS and NLOS paths.

has two headlights and two taillights, a 2 × 2 MIMO link can be considered. The LOS and NLOS paths are considered in this book [17].

2.3.1 The V2V–VLC system considerations

LEDs at the transmitter side and APDs at the receiver side are considered. The market-weighted head-lighting pattern source for the transmitter and receiver [18] is used in the simulation. These types of lights provide good road illumination, as well as very good anti-glare properties [19]. Moreover, practical outdoor VLC links are considered in this book including parked and passing vehicles acting as reflectors.

Each reflector acts as a receiver when the signal is received from the transmitting vehicle and subsequently as a transmitter when the signal is reflected to the receiving vehicle. Because the reflectance characteristics depend on the reflector's nature and physical state, and it also changes with weather conditions, a bidirectional reflectance distribution function (BRDF) is used to model the reflectors in this book. A BRDF model can effectively model the polarimetric signatures of the object surface such as paints or metal [20]. The BRDF is given by [20]

$$f(\theta_{\text{inc}}, \phi_{\text{inc}}, \theta_{\text{ref}}, \phi_{\text{ref}}, \lambda) = \frac{dL_r(\theta_{\text{ref}}, \phi_{\text{ref}})}{dE_i(\theta_{\text{inc}}, \phi_{\text{inc}})} \tag{2.8}$$

where θ_{inc}, ϕ_{inc} respectively denote the incident polar and azimuth angles, θ_{ref}, ϕ_{ref}, respectively, denote the reflected polar and azimuth angles, λ is the wavelength, L_r is the irradiance scattering from the surface, and E_i is the irradiance incident onto the surface.

As outdoor VLC channels undergo fast change effects, a frequency selective channel model is considered for modeling the system [1].

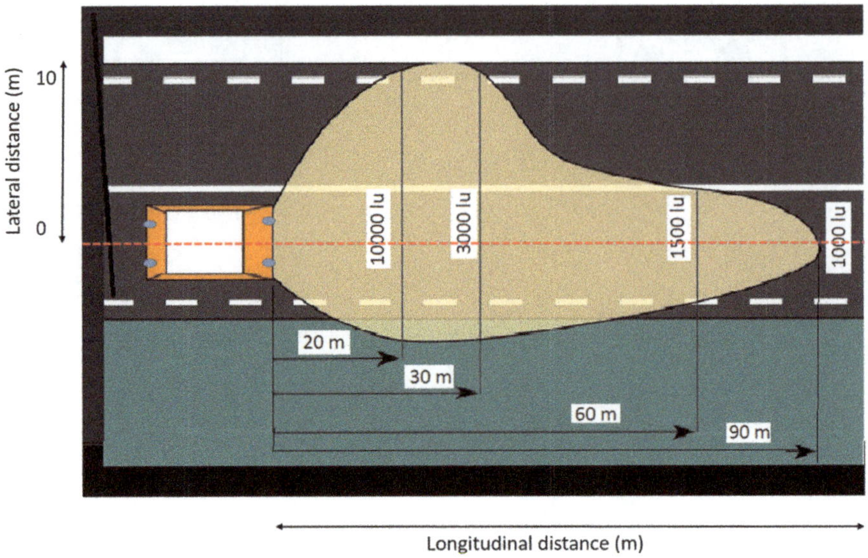

Figure 2.4. Radiation pattern of a vehicle headlight.

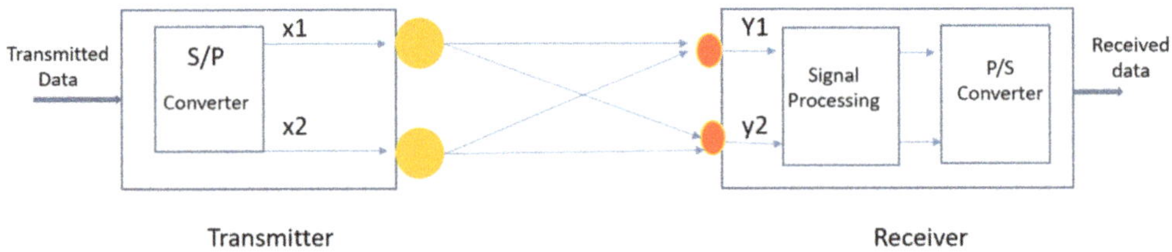

Figure 2.5. Block diagram of a 2 × 2 MIMO link of a V2V–VLC system.

Referring to the road regulations [21], and by using the data provided in [18], and the headlight radiation pattern shown in figure 2.4, the following points should be considered in V2V–VLC system design

1. For the effective luminous intensity of 1000 candela (cd), the road width should be about 10 m and the longitudinal distance should be ranging from 80 to 100 m.
2. The safe distance between vehicles traveling at a speed of 80 km/h, and moving in the same lane must be at least 23 m.
3. The safe distance between vehicles in parallel lanes must be at least 3 m.

Due to the above regulations, the V2V–VLC model can have up to six reflecting vehicles within the testing area and only one LOS path.

2.3.2 System model

The block diagram of the V2V–VLC system is shown in figure 2.5. The transmitted serial data stream of the 2 × 2 MIMO model is converted into two parallel streams using a serial-to-parallel (S/P) converter. Each data stream is intensity modulated and transmitted through LED headlights. At the receiver taillights, PDs convert the light into electrical signals using direct detection.

The received signal can be written as

$$y_r(t) = \gamma x(t) \otimes h(t) + n(t) \tag{2.9}$$

where $y_r(t)$ is the received signal. In vector notation (2.9) is given by

$$\mathbf{y_r} = \gamma \mathbf{H}\,\mathbf{x} + \mathbf{n} \tag{2.10}$$

where $\mathbf{x}$ and $\mathbf{y_r}$ are transmitted and received vectors, respectively, $\mathbf{H}$ is the channel matrix which consists of the LOS and the NLOS components, and $\mathbf{n}$ is the noise vector.

The channel matrix is given by

$$\mathbf{H} = \begin{bmatrix} h_{11} & h_{12} \\ h_{21} & h_{22} \end{bmatrix} \tag{2.11}$$

Referring to figure 2.5, x_1 and x_2 are the transmitted signals from first and second transmitters s_1 and s_2, respectively, y_1 and y_2 are the received signals by both receivers r_1 and r_2, while n_1, and n_2 represent the Gaussian noise at receiver r_1, and r_2, respectively.

2.4 Channel impulse response and transfer function

The total CIR of the V2V–VLC model, $h(t)$ is given by

$$h(t) = h(t)_{\text{LOS}} + h(t)_{\text{NLOS}} \tag{2.12}$$

By performing fast Fourier transform (FFT) on (2.12) we get

$$H(\omega) = H(\omega)_{\text{LOS}} + H(\omega)_{\text{NLOS}} \tag{2.13}$$

2.4.1 Impulse response and transfer function of the LOS component

The 2×2 impulse response matrix $\mathbf{H}_{\text{LOS}}$ between the transmitters and the receivers for the LOS path can be written as

$$\mathbf{H}_{\text{LOS}} = \begin{bmatrix} h_{11} & h_{12} \\ h_{21} & h_{22} \end{bmatrix} \tag{2.14}$$

Any element in $\mathbf{H}_{\text{LOS}}$ is expressed as

$$h_{rs}(t) = G_{rs}\, \delta\, (t - \tau_{rs}) \quad r = 1, 2, \textit{and } s = 1, 2. \tag{2.15}$$

where G_{rs} denotes the gain factor accounting for the losses and gains between the source s and the receiver r, $\tau_{rs} = R_{rs}/c$, denotes the time delay, R_{rs} denotes the distance between the source s and receiver r, c is the speed of light, and $\delta()$ denotes the impulse function.

The gain factor, G_{rs}, is given by [22]

$$G_{rs} = \frac{\cos(\phi_{rs}) \cos(\theta_{rs}) A_r T_s(\theta_{rs}) g(\theta_{rs})}{\pi R_{rs}^2}. \tag{2.16}$$

where ϕ_{rs} denotes the emitting angle between source s and receiver r, θ_{rs} denotes the incident angle between source s and receiver r, A_r denotes the effective area of the receiver, $T_s(\theta_{rs})$ denotes the signal transmission coefficient of an optical filter, and $g(\theta_{rs})$ denotes the concentrator gain. Figure 2.6 shows the relevant angles for the transmitter, the receiver, and the reflector.

In the frequency domain, the LOS transfer function of (2.15) is given by

$$H_{rs}(\omega) = G_{rs}\, e^{-j\omega\tau_{rs}} \tag{2.17}$$

2.4.2 Impulse response and transfer function of the NLOS component

Let each reflector surface be divided into N small surfaces (subsurfaces) elements numbered from $(1, 2, \ldots, N)$ with equal areas A_i and equal reflectivity factors $\rho_i < 1$ $(i = 1, 2, \ldots, N)$, as shown in figure 2.7.

When the transmitted light incidents on the reflector surface, the surface element will be considered as a receiver. Thus, the impulse response matrix $\mathbf{A}$ between the transmitters $(s = 1, 2)$ and the subsurfaces of any reflector, and as shown in figure 2.7, is given by

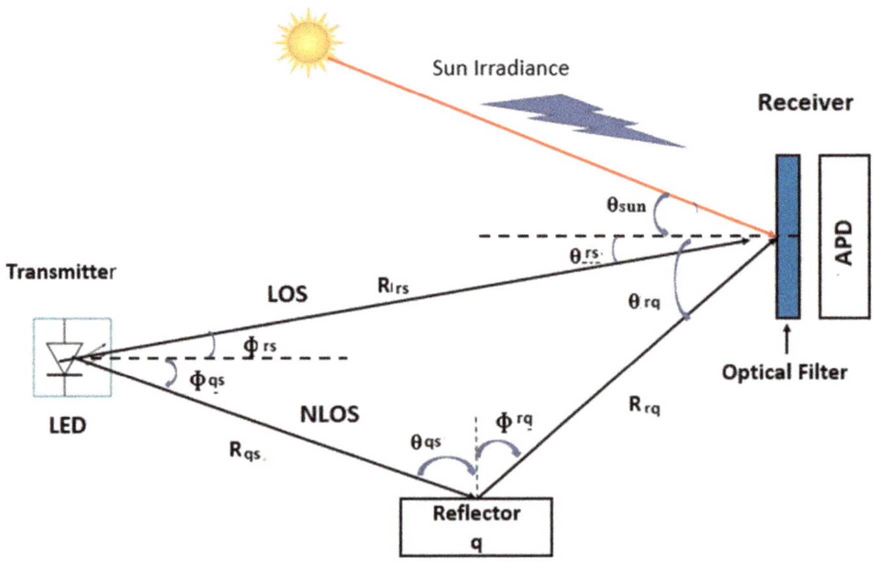

Figure 2.6. Emission and incident angles and other related parameters of an OW system.

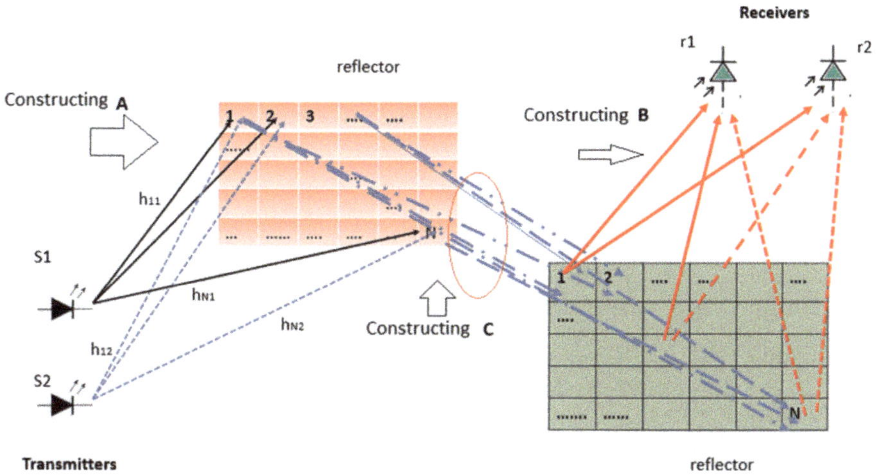

Figure 2.7. Illustration of the NLOS path's matrices (**A, B**, and **C**) construction.

$$\mathbf{A} = \begin{bmatrix} h_{11} & h_{12} \\ h_{21} & h_{22} \\ \cdots & \cdots \\ h_{N1} & h_{N2} \end{bmatrix} \tag{2.18}$$

Each element h_{is} in the matrix **A** is given by

$$h_{is}(t) = G_{is}\,\delta\,(t - \tau_{is}) \quad s = 1, 2, \textit{and } i = 1, 2, \ldots, N \tag{2.19}$$

The gain factor between the transmitter and any subsurface G_{is} is given by [22]

$$G_{is} = \frac{\cos(\phi_{is})\cos(\theta_{is})A_i}{\pi\, R_{is}^2} \tag{2.20}$$

where ϕ_{is} denotes the emitting angle between source s and reflector subsurface i, θ_{is} denotes the incident angle between source s and reflector subsurface i, R_{is} denotes the distance between source s and reflector subsurface i, and A_i denotes the effective area of the reflector subsurface i.

The transfer function of (2.19) $H_{is}(\omega)$ is given by (2.17), with r replaced by i.

Assuming a high order reflection system, i.e. the transmitter s emits the light to reflector 1, which then reflects it to reflector 2 and from 2 to 3 and so on. This process repeats until the last reflector radiates the power to the receiver r directly. In every reflection, the optical power is reduced by the reflectivity factor ρ_i.

The impulse response matrix **B** between the receiver r ($r = 1, 2$), and the last reflector (that has N subsurfaces) is given by

$$\mathbf{B} = \begin{bmatrix} h_{11} & h_{12} & \ldots & h_{1N} \\ h_{21} & h_{22} & \ldots & h_{2N} \end{bmatrix} \tag{2.21}$$

The gain factor G_{ri}, and the transfer function $H_{ri}(\omega)$ of each element in (2.21) are given by (2.15) and (2.17), respectively, but with the proper replacements of i instead of s.

The corresponding impulse response matrix between any two reflectors (each with N subsurfaces) $\mathbf{C}$ is

$$\mathbf{C} = \begin{bmatrix} h_{11} & h_{12} & \dots & \dots & h_{1N} \\ h_{21} & h_{22} & \dots & \dots & h_{2N} \\ \dots & \dots & \dots & \dots & \dots \\ h_{N1} & h_{N2} & \dots & \dots & h_{NN} \end{bmatrix} \tag{2.22}$$

where h_{ki}, and the gain G_{ki}, are given by (2.15) and (2.16), respectively, $(i, k = 1, 2, \dots, N$, (i and k) are the subsurface's index of the first and second reflector, respectively. Also, the transfer function $H_{ki}(\omega)$ is given by (2.17), after proper replacement of k with r, and i with s. Figure 2.7 illustrates how the NLOS path's matrices (**A, B**, and **C**) are constructed.

The total impulse response $h_{\text{NLOS}}(t)$ for the NLOS components is the sum of the contributions from the reflections up to a given order L [23]

$$h_{\text{NLOS}}(t) \approx \sum_{l=1}^{L} h_{\text{NLOS}}^{(l)}(t) \tag{2.23}$$

As an example, the first order reflection system (i.e. one reflector with N subsurfaces ($i = 1, 2, \dots, N$) has an impulse response given by [22]

$$h_{\text{NLOS}}^{(1)}(t) = \sum_{i=1}^{N} \rho_i h_{ri}(t) \otimes h_{is}(t) \tag{2.24}$$

The second order reflection system (i.e. two reflectors with N subsurfaces each $(i, k = 1, 2, \dots, N)$ is given by

$$h_{\text{NLOS}}^{(2)}(t) = \sum_{i=1}^{N} \sum_{k=1}^{N} \rho_k \rho_i h_{kr}(t) \otimes h_{ki}(t) \otimes h_{is}(t) \tag{2.25}$$

The third order reflection system (i.e. three reflectors with N subsurfaces each $(i, k, l = 1, 2, \dots, N)$, ($l$ is the subsurface's index of the third reflector) is given by

$$h_{\text{NLOS}}^{(3)}(t) = \sum_{l=1}^{N} \sum_{i=1}^{N} \sum_{k=1}^{N} \rho_k \rho_i \ \rho_l h_{kr}(t) \otimes h_{li}(t) \otimes h_{kl}(t) \otimes h_{is}(t) \tag{2.26}$$

and so on.

Many previous works such as [24] and [25] used recursive and iterative formulas to approximate (2.23). By the end, they got an exponential decay model for the VLC outdoor environment which is given by

$$h_{\text{NLOS}}(t, \tau) = \frac{1}{\tau} e^{\frac{-t}{\tau}} \tag{2.27}$$

where the time constant $\tau = 2\ \tau_{\text{RMS}}$ and τ_{RMS} is the channel RMS delay spread.

2.4.3 Derivation of NLOS transfer function

The frequency domain version of (2.18) is given by

$$\mathbf{A}(\omega) = [\mathbf{h}_{i1}(\omega) \quad \mathbf{h}_{i2}(\omega)] \tag{2.28}$$

where $\mathbf{h}_{i1}(\omega)$ and $\mathbf{h}_{i2}(\omega)$ are two vectors that indicate the transfer functions of the first and second transmitter and the reflector subsurfaces, respectively.

Any vectors in (2.28) can be expanded to

$$\mathbf{h}_{is}(\omega) = [G_{1s}\, e^{-j\omega\tau_{1s}}\; G_{2s}\, e^{-j\omega\tau_{2s}} \ldots G_{Ns}\, e^{-j\omega\tau_{Ns}})]^T \quad s = 1, 2 \tag{2.29}$$

Similarly, the frequency domain version of (2.21) is given by

$$\mathbf{B}(\omega)^T = [\mathbf{h}_{1i}(\omega) \quad \mathbf{h}_{2i}(\omega))] \tag{2.30}$$

where $\mathbf{h}_{ri}(\omega)$ and $\mathbf{h}_{2i}(\omega)$ are two vectors that indicate the transfer functions of the first and second receiver with the last reflector surface.

The vectors in (2.30) can be expanded to

$$\mathbf{h}_{ri}(\omega)^T = [G_{r1}\, e^{-j\omega\tau_{r1}}\; G_{r2}\, e^{-j\omega\tau_{r2}} \ldots G_{rN}\, e^{-j\omega\tau_{rN}})]^T \quad r = 1, 2 \tag{2.31}$$

The $N \times N$ frequency domain version of (2.22) can be written as

$$\mathbf{C}(\omega) = \begin{bmatrix} G_{11}\, e^{-j\omega\tau_{11}} & G_{12}\, e^{-j\omega\tau_{12}} & \ldots & G_{1N}\, e^{-j\omega\tau_{1N}} \\ G_{21}\, e^{-j\omega\tau_{21}} & G_{22}\, e^{-j\omega\tau_{22}} & \ldots & G_{2N}\, e^{-j\omega\tau_{2N}} \\ \ldots & \ldots & \ldots & \ldots \\ G_{N1}\, e^{-j\omega\tau_{N2}} & G_{N2}\, e^{-j\omega\tau_{N2}} & \ldots & G_{NN}\, e^{-j\omega\tau_{NN}} \end{bmatrix} \tag{2.32}$$

To include the reflectivity factors ρ_i of the surfaces, we define the $N \times N$ reflectivity matrix as follows

$$\mathbf{G}_\rho = \mathrm{diag}[\rho_1\, \rho_2 \ldots \rho_N] \tag{2.33}$$

If all the reflection coefficients are assumed to be equal i.e, $(\rho_1 = \rho_2 = \cdots = \rho_N)$.

To have a closed form expression for the NLOS link transfer function, we have to find the FFT of (2.24), (2.25) and (2.26) till lth reflection order, then applying Carl Neumann's result on matrices as follows, for first order reflection

$$\mathbf{H}^{(1)}{}_{\mathrm{NLOS}}(\omega) = \mathbf{B}(\omega)\, \mathbf{G}_\rho\, \mathbf{A}(\omega) \tag{2.34}$$

For second-order reflection

$$\mathbf{H}^{(2)}{}_{\mathrm{NLOS}}(\omega) = \mathbf{B}(\omega)\, \mathbf{G}_\rho\, \mathbf{C}(\omega)\, \mathbf{G}_\rho\, \mathbf{A}(\omega) \tag{2.35}$$

For third order reflection

$$\mathbf{H}^{(3)}{}_{\mathrm{NLOS}}(\omega) = \mathbf{B}(\omega)\, \mathbf{G}_\rho\, (\mathbf{C}(\omega)\, \mathbf{G}_\rho)^2\, \mathbf{A}(\omega) \tag{2.36}$$

For the lth order reflection

$$\mathbf{H}^{(l)}{}_{\mathrm{NLOS}}(\omega) = \mathbf{B}(\omega)\,\mathbf{G}_{\rho}\,(\mathbf{C}(\omega)\,\mathbf{G}_{\rho})^{l-1}\,\mathbf{A}(\omega) \tag{2.37}$$

The total NLOS transfer function is given by the infinite series

$$\begin{aligned}
\mathbf{H}_{\mathrm{NLOS}}(\omega) &= \sum_{l=1}^{\infty} \mathbf{H}^{(l)}_{\mathrm{NLOS}}(\omega) \\
&= \mathbf{H}^{(1)}{}_{\mathrm{NLOS}}(\omega) + \mathbf{H}^{(2)}{}_{\mathrm{NLOS}}(\omega) + \mathbf{H}^{(3)}{}_{\mathrm{NLOS}}(\omega) + \cdots + \mathbf{H}^{(l)}{}_{\mathrm{NLOS}}(\omega) + \cdots \\
&= \mathbf{B}(\omega)\,\mathbf{G}_{\rho}\,\mathbf{A}(\omega) + \mathbf{B}(\omega)\,\mathbf{G}_{\rho}\,\mathbf{C}(\omega)\,\mathbf{G}_{\rho}\,\mathbf{A}(\omega) \\
&\quad + \mathbf{B}(\omega)\,\mathbf{G}_{\rho}\,(\mathbf{C}(\omega)\,\mathbf{G}_{\rho})^{2}\,\mathbf{A}(\omega) + \cdots \\
&\quad + \mathbf{B}(\omega)\,\mathbf{G}_{\rho}\,(\mathbf{C}(\omega)\,\mathbf{G}_{\rho})^{l-1}\,\mathbf{A}(\omega) + \cdots \\
&= \mathbf{B}(\omega)\,\mathbf{G}_{\rho}\,\sum_{m=0}^{\infty}(\mathbf{C}(\omega)\,\mathbf{G}_{\rho})^{m}\,\mathbf{A}(\omega)
\end{aligned} \tag{2.38}$$

Applying Carl Neumann's result on matrices ($(\mathbf{I}_n - \mathbf{F})^{-1} = \sum_{m=0}^{\infty}\mathbf{F}^{m}$), then, the expression in (2.38) can be summed up to

$$\sum_{m=0}^{\infty}(\mathbf{C}(\omega)\,\mathbf{G}_{\rho})^{m} = (\mathbf{I} - \mathbf{C}(\omega)\,\mathbf{G}_{\rho})^{-1} \tag{2.39}$$

where $\mathbf{I}$ denotes the unity matrix.

Rewrite (2.38) to get the final expression for the NLOS transfer function which is given by the matrix product as

$$\mathbf{H}_{\mathrm{NLOS}}(\omega) = \mathbf{B}(\omega)\,\mathbf{G}_{\rho}\,(\mathbf{I} - \mathbf{C}(\omega)\,\mathbf{G}_{\rho})^{-1}\,\mathbf{A}(\omega) \tag{2.40}$$

Equation (2.40) gives a closed form expression for the NLOS path, where the infinite number of reflections shown by (2.38) becomes not an issue.

The impulse response $h_{\mathrm{NLOS}}(t)$ can then be obtained by performing inverse fast Fourier transform (IFFT) on the transfer function $\mathbf{H}_{\mathrm{NLOS}}(\omega)$. Therefore, the total impulse response of the suggested model $h(t)$ can then be obtained by performing IFFT on the transfer function given by (2.13).

2.5 Performance analysis of the V2V–VLC channel model

2.5.1 The received optical power and SNR

The total received optical power P_r can be calculated as follows:

$$P_r = P_r^{\mathrm{LOS}} + P_r^{\mathrm{NLOS}} \tag{2.41}$$

The received optical power from the LOS path is given by

$$P_r^{\text{LOS}} = \begin{cases} \left(\sum_{r=1}^{2} \sum_{s=1}^{2} \frac{I_s(\alpha_s, \beta_s) A_r}{(\text{LER})\, R_{rs}^2} \cos(\phi_{rs}) \cos(\theta_{rs})\right) & \text{if } 0 \leqslant \theta_{rs} \leqslant \Psi_{rs} \\ 0, & \text{if } \theta_{rs} > \Psi_{rs} \end{cases} \tag{2.42}$$

where $I_s(\alpha_s, \beta_s)$ is the luminous intensity of the transmitter s from the direction (α_s, β_s) and measured in candela (cd), (α_s, β_s) are the vertical and horizontal angles of the headlight, respectively, Ψ is the receiver's FOV, LER is luminous efficacy of radiation of a high power LED.

The received optical power from the NLOS paths is given by

$$P_r^{\text{NLOS}} = \begin{cases} \left(\sum_{r=1}^{2} \sum_{s=1}^{2} \sum_{q=1}^{Q} \frac{I(\alpha_s, \beta_s) \cos(\theta_{qr})\, A_q A_r}{(\text{LER})\, R_{qs}^2 \left(R_{rq}^2 + h_i^2\right)} \rho_q \cos(\phi_{rq}) \cos(\theta_{rq})\right) & \text{if } 0 \leqslant \theta_{rq} \leqslant \Psi_{rq} \\ 0 & \text{if } \theta_{rq} > \Psi_{rq} \end{cases} \tag{2.43}$$

where Q is the total number of the reflectors, θ_{qs} denotes the incident angle between reflector q and receiver r, A_q denotes the reflector effective area, h_i denotes the hight of the receiver, R_{qs} denotes the distance between the transmitter and the reflector, R_{rq} denotes the distance between the reflector and the receiver, ϕ_{rq} denotes the emission angle between the reflector and receiver, θ_{rq} denotes the incident angle between the receiver and the reflector, and ρ_q denotes the reflection factor of reflector q. The illustration of these parameters can be seen in figure 2.6.

In this chapter, we assume the OOK modulation scheme and both transmitters have the same power. Thus, at the receiver, the SNR is given by [24, 26]

$$\text{SNR} = \frac{(\gamma P_r)^2}{\sigma_n^2} \tag{2.44}$$

where σ_n^2 is the noise power given by (1.6).

2.5.2 Calculating the BER

BER is the probability that an error may occur in a bit in the pulse train, i.e., a '1' bit turns into a '0' bit or vice versa.

$$\text{BER} = P(0)\, P(e/0) + P(1)\, P(e/1) \tag{2.45}$$

where $P(0)$ and $P(1)$ are the probability of transmit '0' and '1' respectively, $P(e/0)$ and $P(e/1)$ are the conditional probability for receiving '1' while '0' was transmitted and receiving '0' while '1' was transmitted, respectively.

OOK-NRZ transmission is considered and random noise Gaussian distribution. Moreover, Let both symbols have same variances and they have the same probability

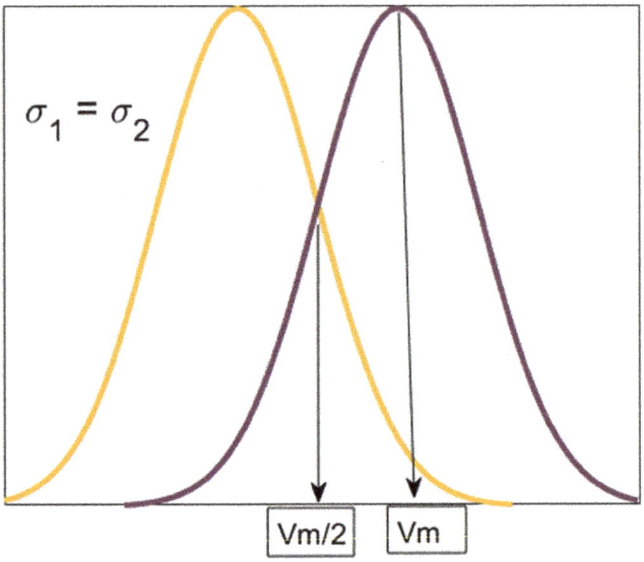

Figure 2.8. Conditional probability density function with OOK modulation.

to be transmitted (equally likely), i.e. $P(0) = P(1) = 1/2$, in this case, the threshold level is set to be the halfway between the symbols, i.e at $V_m/2$ as shown in figure 2.8.

The conditional probability functions can be calculated as

$$P(e/1) = \frac{1}{\sigma\sqrt{2\pi}} \int_{-\infty}^{\frac{V_m}{2}} e^{\frac{-(x-\mu_1)^2}{2\sigma^2}} dx \tag{2.46}$$

$$P(e/0) = \frac{1}{\sigma\sqrt{2\pi}} \int_{\frac{V_m}{2}}^{\infty} e^{\frac{-(x-\mu_0)^2}{2\sigma^2}} dx \tag{2.47}$$

where $\mu_0 = 0$ and $\mu_1 = V_m$ for NRZ signal.

substituting (2.46) and (2.47) in (2.45)

$$\begin{aligned} \mathrm{BER_{OOK}} &= \frac{1}{2\sigma\sqrt{2\pi}} \left(\int_{-\infty}^{\frac{V_m}{2}} e^{\frac{-(x-V_m)^2}{2\sigma^2}} dx + \int_{\frac{V_m}{2}}^{\infty} e^{\frac{-(x-0)^2}{2\sigma^2}} dx \right) \\ &= \frac{1}{2}\left(\frac{1}{2} erfc\left(\frac{1}{2\sqrt{2}} \sqrt{\mathrm{SNR}} \right) + \frac{1}{2} erfc\left(\frac{1}{2\sqrt{2}} \sqrt{\mathrm{SNR}} \right) \right) \\ &= \frac{1}{2} erfc\left(\frac{1}{2\sqrt{2}} \sqrt{\mathrm{SNR}} \right) \\ &= \frac{1}{2} erfc\left(\frac{1}{2\sqrt{2}} \sqrt{\frac{(\gamma P_r)^2}{\sigma_n^2}} \right) \end{aligned} \tag{2.48}$$

where $erfc(.)$ is the complementary error function used to calculate the tail probability of the standard Gaussian distribution, given by

$$erfc(x) = \frac{2}{\sqrt[2]{\pi}} \int_{x}^{\infty} e^{-y^2}\, \mathrm{d}y \tag{2.49}$$

2.5.3 Time dispersion parameters for the channel

By using the derived CIR given by (2.12), we can calculate several channel parameters such as RMS delay spread, mean excess delay, frequency correlation function, coherence time, and coherence bandwidth.

The RMS delay spread and mean excess delay are given by [27]

$$\tau_{\mathrm{RMS}} = \sqrt{\frac{\int_0^{\infty} (t - \bar{\tau})^2\, h^2(t)dt}{\int_0^{\infty} h^2(t)dt}} \tag{2.50}$$

where $\bar{\tau}$ is the mean excess delay and is given by

$$\bar{\tau} = \frac{\int_0^{\infty} t\, h^2(t)dt}{\int_0^{\infty} h^2(t)dt} \tag{2.51}$$

For V2V–VLC environment, the frequency correlation function $H(\triangle\omega)$ can be obtained from the CIR as [28]

$$H(\triangle\omega) = \int_{-\infty}^{\infty} h(t)\, e^{-j\triangle\omega t}\, dt \tag{2.52}$$

Once the frequency correlation function is calculated, the coherence bandwidth B_c for a VLC channel can be found as [29]

$$B_c = |\min\triangle(\omega)| \tag{2.53}$$

Since the V2V–VLC system is good for short distance communications, the effective velocity of both the transmitter and the receiver $\left(V_{\mathrm{eff}} = \sqrt{V_s^2 + V_r^2}\right)$ compared to the velocity of light is very small, thus, a correlation level of 0.9 is typically chosen [29], i.e $|H\triangle(\omega)| = 0.9$.

2.6 Simulation and results

MATLAB software is used to simulate the V2V–VLC system. The simulation parameters are shown in tables 1.3 and 2.1. Multi-path propagation is considered. We assume all reflector surfaces to have the same reflectively factor and the same areas. We consider the moving speed of 30 km/h for the transmitter and 25 km/h for the receiver(whether they move in the same direction, or in the opposite direction). We use Philips Ultinon LED 12985BWX2 as transmitter [30]. Also, we use APD model (Si) Hamamatsu S8664 – 1010 as a receiver [31]. We analyze the system performance considering various parameters.

Figure 2.9 shows the CIR of the LOS and the NLOS components when the transmitter and the receiver are moving in the same direction, but with different velocities. The CIR of the LOS path is represented by an impulse, while the CIR of

Table 2.1. Simulation parameters.

Parameter	Value
Number of reflectors	2–6
Modulation type	OOK
Distance between the reflectors	3–80 m
Distance between the transmitter and the receiver	20–120 m
Receiver area	4 cm^2
Total area of the reflector	0.5 m^2
LED luminance intensity	1000–2500 *cd*
Transmitted and received angles	5°–45°
Responsivity of PD γ	0.54 A/W
Reflectivity ρ	0.8
Gain of the optical filter T (Ψ)	1
Gain of the optical concentrator g	5
Frequency correlation level	0.9
Speeds of the transmitter and receiver, respectively	30 km/h, 25 km/h
Data rate	35 Mbps
Receiver hight	70 cm

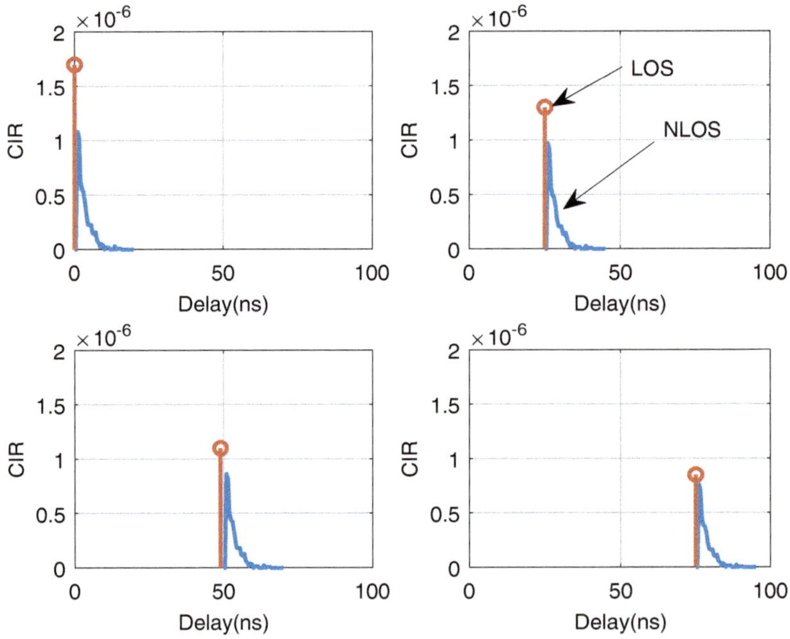

Figure 2.9. Channel impulse response of the LOS and NLOS components when vehicles are moving in the same direction.

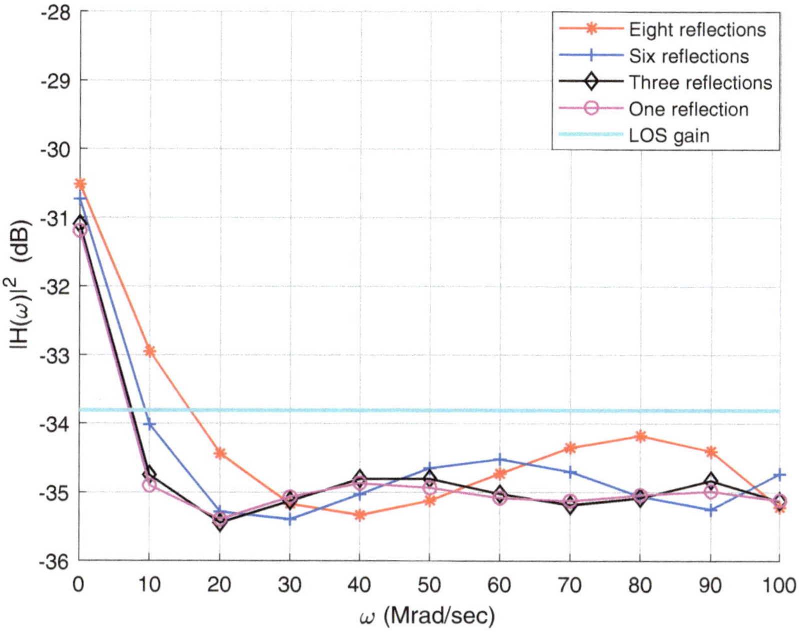

Figure 2.10. Transfer functions of the LOS and NLOS VLC links.

the NLOS components has an exponential decay behavior. Moreover, figure 2.9 shows that, as the distance between the transmitter and the receiver increases, the delay increases but the channel gain decreases. The low gain shown in figure 2.9 for both paths is due to the nature of the V2V–VLC system where the channel is changing very fast. In addition, both the transmitter and the receiver are moving affecting both the angle of emitting, as well as, the angles of incidence. The importance of the channel gain in the V2V–VLC system rising from it determines the achievable SNR for fixed transmitting power. Figure 2.10 shows the transfer function of the V2V–VLC channel. We use (2.40) to simulate the NLOS paths for up to six reflectors, while we use (2.16) and (2.17) to simulate the gain of the LOS path. From figure 2.10, we see that the gain of the channel depends not only on frequency but also on other factors such as emitting angle, incident angle, and the reflectivity of the surrounding objects. Also, figure 2.10 shows the difference between LOS gain and NLOS gain is inversely proportional to a number of reflections. This means that we need a high number of reflections to make the difference close to zero. The exponential decay of the CIR for the diffusion path shown in figure 2.9 is obtained after performing IFFT for the transfer function shown in figure 2.10. A smoothing operation has to be done for the transfer function (such as raised cosine window) before applying the IFFT. The simulation results show acceptable similarity with the mathematical model for CIR given by (2.27) as shown in figure 2.11.

Figure 2.12 shows the received power versus the distance between the vehicles for up to six reflectors. The received power is inversely proportional to the distance. Furthermore, it can be seen that the contribution by the high order reflections ($l > 2$)

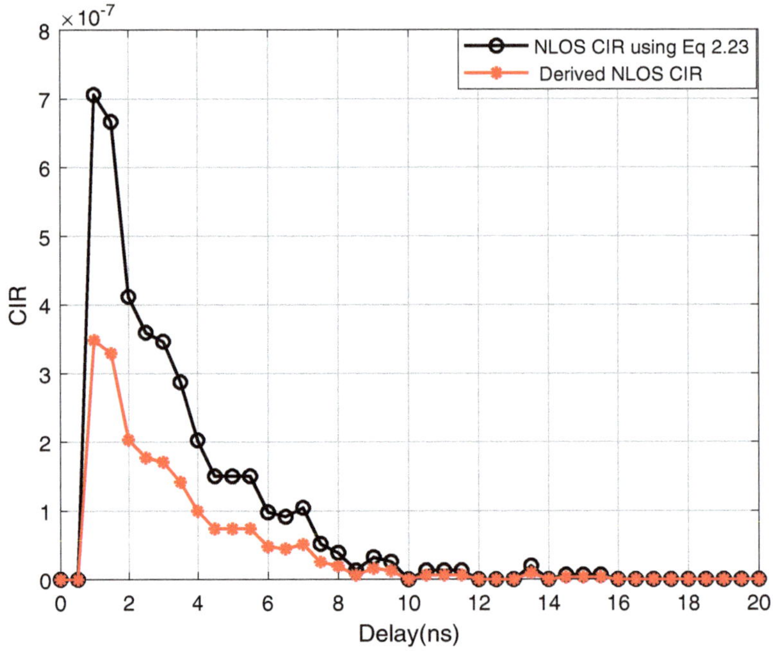

Figure 2.11. NLOS channel impulse response, derived and obtained using EQ (2.23).

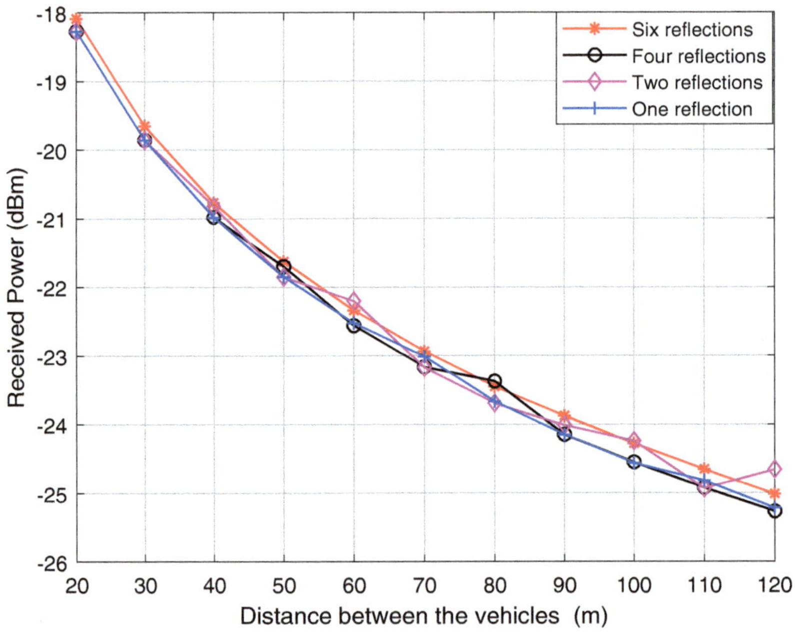

Figure 2.12. Received power versus the distance between vehicles.

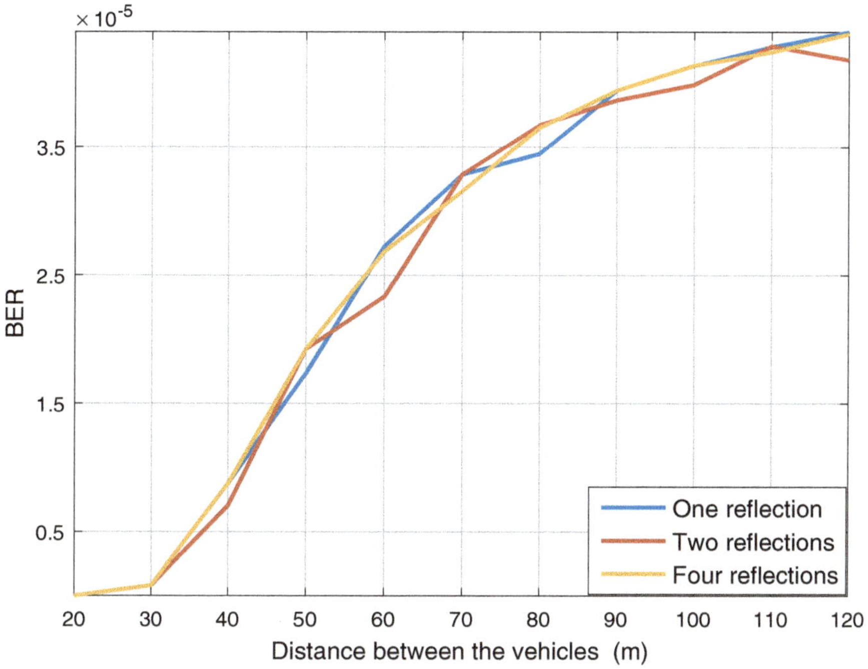

Figure 2.13. BER versus the distance when both vehicles move in the opposite direction.

are not dominant. The dominant components are coming from the LOS path as well as from the first and second reflections. This is because of the scattering property of the light and the absorption by the reflectors and other surrounding objects reduce the received power strength. BER versus different values for R_{rs} at a data rate of 35 Mbps is shown in figure 2.13. It can be seen that, as the distance increases, the system BER performance decreases. Also, we can see from figure 2.13 that the value of the BER degrades from 10^{-6} at R_{rs} = 30 m to 3.5×10^{-5} at R_{rs} = 80 m. This happens because when R_{rs} changes from 30 to 80 m, the received optical power reduces by more than three times (e.g., it decreases from −24 dBm to −20 dBm) according to figure 2.12.

The investigated model shows better performance than the results obtained in [9] and [10], since we consider passing by and parking cars as reflectors. Also, figure 2.13 emphasizes that the high order number of reflections will not cause significant difference in the performance of the system.

Figure 2.14 shows the PDF of the delay spread for the V2V–VLC channel. It can be seen that the best fitted distribution for the RMS of the channel in the case of the same direction movement is similar to Rician distribution. The mean value of the delay spread of V2V–VLC model is about 12 ns, while the authors in [32] found it to be 10–15 ns, as shown in table 2.2. This is attributed to the fact that the distance is short while the signal speed is very high in V2V–VLC systems, which produce a dispersive channel. In addition, motion introduces additional Doppler spreading and shifting.

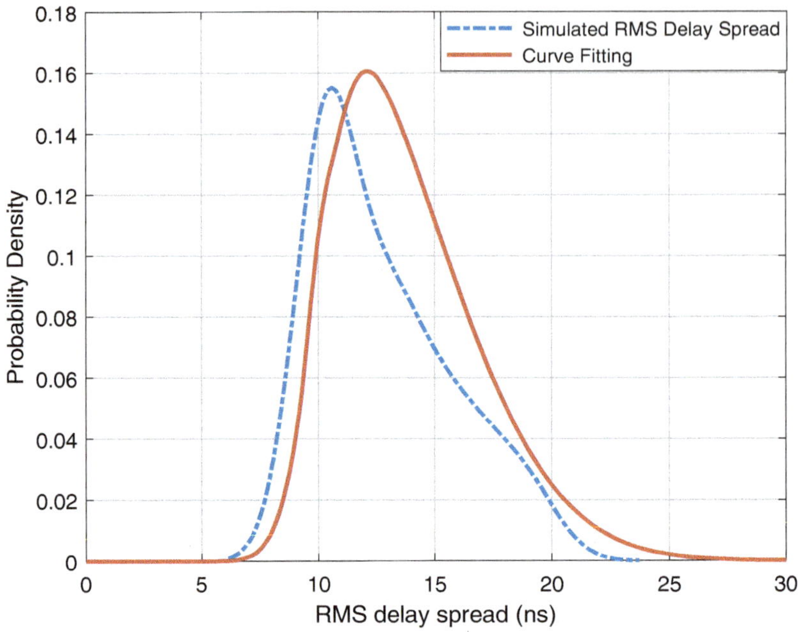

Figure 2.14. RMS delay spread for one reflection system.

Table 2.2. RMS delay spread comparison.

Reference	Delay spread (ns)
V2V [32]	10
V2I [32]	15
Infrared channels [4]	2
For this work both vehicles move in the same direction	12

Figures 2.15 and 2.16 depict the average of 50% and 90% coherence time of the suggested channel when the transmitter and the receiver are moving in the same direction, and in the opposite direction, respectively. There is a decreasing trend in coherence time with effective speed for both directions, as expected, which is consistent with the Doppler spread analysis. The VLC channel shows a larger coherence time than the values reported for a V2V–RF channel in the urban scenarios [33, 34]. Table 2.3 compares the median coherence time of (VLC and RF) obtained from this work and several previous works. Moreover, the V2V–VLC model shows that 2×2 MIMO-VLC links have much slower channel time variation compared to RF V2V links, and other VLC models. Also, the V2V–VLC channel coherence time is found to be at least an order of magnitude larger than that of RF channels, indicating that the V2V–VLC channels are very stable.

Here, the coherence time was computed directly from received signals instead of assuming a reciprocal time–frequency relationship. The fact that the coherence time

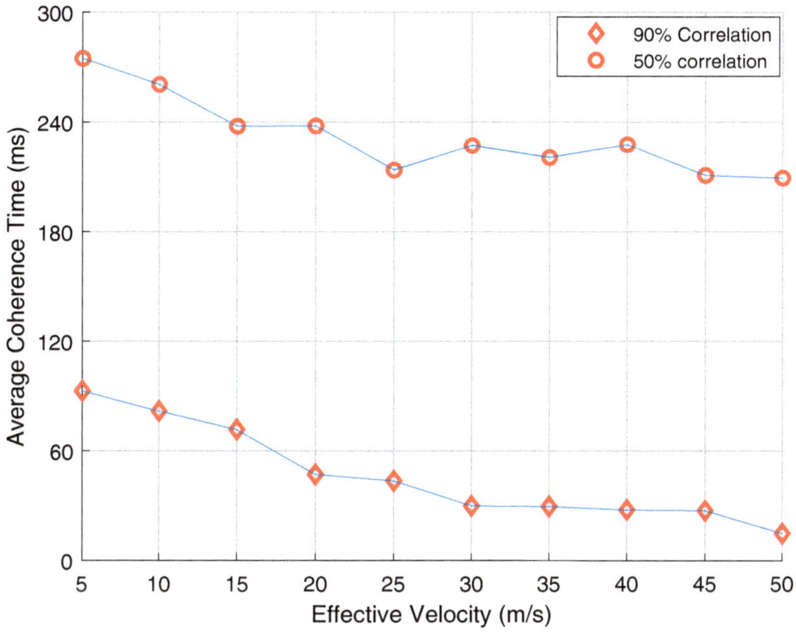

Figure 2.15. Channel coherence time when both vehicles move in the same direction.

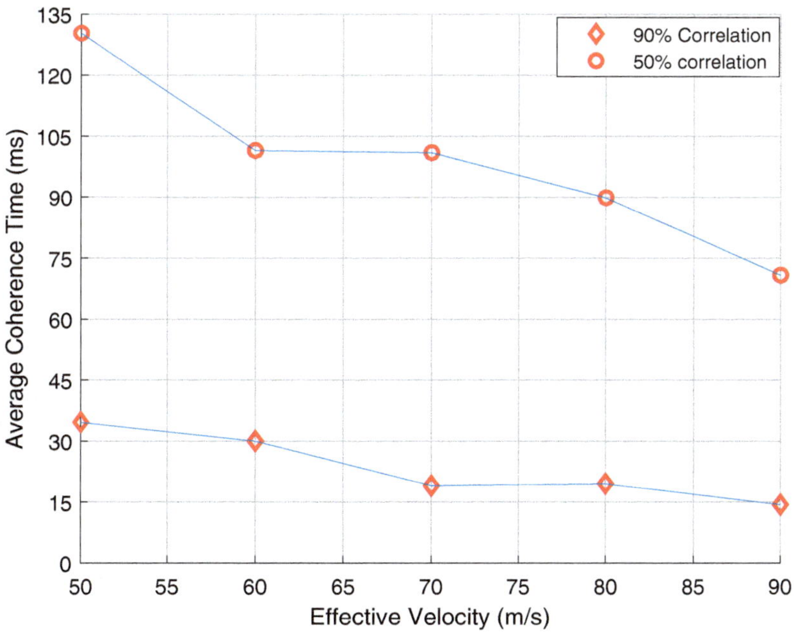

Figure 2.16. Channel coherence time when both vehicles move in opposite directions.

Table 2.3. Median coherence time comparison (units in ms).

Channel	50%	90%
RF channel [34]	103	3.6
VLC channel [35]	164.7	46.8
VLC channel [12]	367	33
For this work both vehicles move in the same direction	220	50
For this work both vehicles move in the opposite direction	100	25

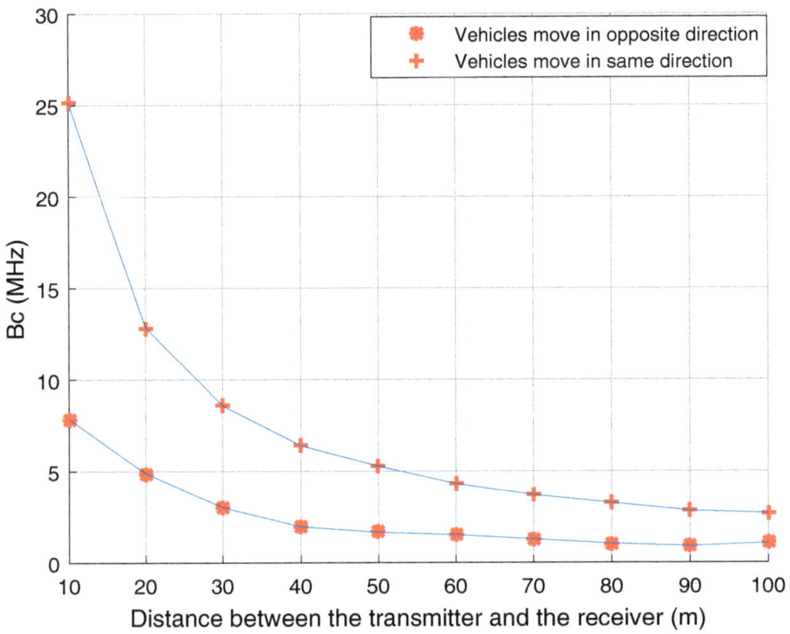

Figure 2.17. Coherence bandwidth for one reflection system with 90% correlation when vehicles move in the same and opposite directions.

of the VLC system is high, especially when both vehicles move in the same direction, is attributed to the very high frequency of light. The movement in the opposite direction appears to have lower channel coherence time than the movement in the same direction. This is because the effective speed of both vehicles will be higher, and the dominant path is the NLOS only. We can suggest that the V2V–VLC channel can be considered as a frequency selective channel. Note that the simulated coherence time does not continue to decrease monotonically with distance since VLC signal strength varies rapidly, it is plausible that small perturbations in the environment could lead to relatively large changes in the channel gain.

The coherence bandwidth values as a function of separation distances are shown in figure 2.17, where the instantaneous coherence bandwidth is inversely proportional to the separation distance. The highest coherence bandwidth values are available for the lowest RMS delay spread values, and higher values can be seen for the same direction movement. The coherence bandwidth values range between 3 and 7 MHz for the opposite direction movement, and 4–25 MHz for the same direction movement.

Low RMS delay spread values lead to higher coherence bandwidth because in the same direction of movement, the NLOS path can be considered dominant. Also, this result shows the importance of the scattering environment of the channel, as the paths take less time to reach the receiver, so the coherence bandwidth of the channel increases.

2.7 Chapter summary

In this chapter, we model the V2V–VLC communication system as 2×2 MIMO, considering the real-time environment, practical reflectors, and real ambient noise. Then, the channel of communication is modeled considering LOS and NLOS paths based on upcoming IEEE 802.15.7r1. We derive the CIR expressions for LOS and NLOS paths. Finally, we calculate the channel parameters such as BER, SNR, RMS delay spread, coherence bandwidth, and coherence time when the vehicles move in the same direction and when they move in the opposite direction.

References

[1] Uysal M, Ghassemlooy Z, Bekkali A, Kadri A and Menouar H 2015 Visible light communication for vehicular networking: Performance study of a V2V system using a measured headlamp beam pattern model *IEEE Veh. Technol. Mag.* **10** 45–53

[2] Popoola W, Ghassemlooy Z and Rajbhandari S 2013 *Optical Wireless Communications System and Channel Modelling with MATLAB* (Boca Raton, FL: CRC Press, Taylor and Francis Group)

[3] Carruthers J B and Kahn J M 1996 Multiple-subcarrier modulation for nondirected wireless infrared communication *IEEE J. Sel. Areas Commun.* **14** 538–46

[4] Carruthers J B and Kahn J M 1997 Modeling of nondirected wireless infrared channels *IEEE Trans. Commun.* **45** 1260–8

[5] Hayasaka N and Ito T 2007 Channel modeling of nondirected wireless infrared indoor diffuse link *Electron. Commun. Japan (Part I: Commun.)* **90** 9–19

[6] Carruthers J B and Carroll S M 2005 Statistical impulse response models for indoor optical wireless channels *Intl J. Commun. Syst.* **18** 267–84

[7] Kahn J M, Krause W J and Carruthers J B 1995 Experimental characterization of non-directed indoor infrared channels *IEEE Trans. Commun.* **43** 1613–23

[8] Jungnickel V, Nonnig V P S and von Helmolt C 2002 A physical model of the wireless infrared communication channel *IEEE J. Sel. Areas Commun.* **20** 631–40

[9] Luo P, Ghassemlooy Z, Minh H L, Bentley E, Burton A and Tang X 2014 Fundamental analysis of a car to car visible light communication system *9th Int. Symp. on Communication Systems, Networks Digital Signal Processing (CSNDSP) (July 2014)* pp 1011–16

[10] Luo P *et al* 2015 Performance analysis of a car-to-car visible light communication system *Appl. Opt.* **54** 1696–706

[11] Lee I E, Sim M L and Kung F W L 2011 A dual-receiving visible-light communication system under time-variant non-clear sky channel for intelligent transportation system *16th European Conf. on Networks and Optical Communications (NOC)* pp 153–6
[12] Cui Z, Wang C and Tsai H-M 2014 Characterizing channel fading in vehicular visible light communications with video data *Vehicular Networking Conf. (VNC), 2014 IEEE* pp 226–9
[13] Jung S-Y, Lee S J, Kwon J K and Kwon Y-H 2012 Evaluation of visible light communication channel delay profiles for automotive applications *EURASIP J. Wirel. Comm.* **1** 370–9
[14] Araki N and Yashima H 2005 A channel model of optical wireless communications during rainfall *2nd Int. Symp. on Wireless Communication Systems (Sept 2005)* pp 205–9
[15] Lee S J, Kwon J K, Jung S Y and Kwon Y H 2012 Simulation modeling of visible light communication channel for automotive applications *15th Int. IEEE Conf. on Intelligent Transportation Systems (ITSC) (Sept 2012)* pp 463–8
[16] Kinoshita M, Yamazato T, Okada H, Fujii T, Arai S, Yendo T and Kamakura K 2014 Motion modeling of mobile transmitter for image sensor based I2V–VLC, V2I–VLC, and V2V–VLC *Globecom Workshops (GC Wkshps) (Dec 2014)* pp 450–5
[17] Farahneh H, Khalifeh A and Fernando X 2016 An outdoor multi path channel model for vehicular visible light communication systems *Photonics North (PN) (May 2016)* pp 1–1
[18] Kojima S, Sivak M, Flannagan M J and Traube E C 2005 A market weighted description of low-beam headlighting patterns in the us *Proc. of the IEEE* **1** 61–71
[19] Headlamps properties http://en.wikipedia.org/wiki/Headlamp
[20] Pan J, Chen Q, Qian W and Geng L 2015 Results of a new polarimetric BRDF simulation of metallic surfaces *Infrared Phys. Technol.* **72** 58–67
[21] Ministry of Transportation of Canada http://www.mto.gov.on.ca/english/dandv/driver/handbook
[22] Schulze H 2016 Frequency-domain simulation of the indoor wireless optical communication channel *IEEE Trans. Commun.* **64** 2551–62
[23] Barry J R and Kahn J M 1997 Wireless infrared communications *Proc. of the IEEE* **85** 265–98
[24] Lee K, Li C, Yi Y and Lee K 2014 Performance analysis of visible light communication using the STBC-OFDM technique for intelligent transportation systems *Int. J. Electron.* **101** 1117–33
[25] Wang C, Yu H Y and Zhu Y J 2016 A long distance underwater visible light communication system with single photon avalanche diode *IEEE Photonics J.* **8** 1–11
[26] Do T-H and Yoo M 2012 Received power and SNR optimization for visible light communication system *Fourth Int. Conf. on Ubiquitous and Future Networks (ICUFN)*
[27] Sarbazi E, Uysal M, Abdallah M and Qaraqe K 2014 Ray tracing based channel modeling for visible light communications *22nd Signal Processing and Communications Applications Conf. (SIU) (April 2014)* pp 702–5
[28] Miramirkhani F and Uysal M 2015 Channel modeling and characterization for visible light communications *IEEE Photonics J.* **7** 1–16
[29] Varela M S and Sanchez M G 2001 RMS delay and coherence bandwidth measurements in indoor radio channels in the UHF band *IEEE Trans. Veh. Technol.* **50** 515–25
[30] Light emiting diode http://www.powerbulbs.com
[31] Photo diode parameters http://www.hamamatsu.com

[32] Lee S J, Kwon J K, Jung S-Y and Kwon Y-H 2012 Evaluation of visible light communication channel delay profiles for automotive applications *EURASIP J. Wirel. Comm.* 1823–6

[33] Mecklenbrauker C F, Molisch A F, Karedal J, Tufvesson F, Paier A, Bernado L, Zemen T, Klemp O and Czink N 2011 Vehicular channel characterization and its implications for wireless system design and performance *Proc. IEEE* **99** 1189–212

[34] Cheng L, Henty B E, Stancil D D, Bai F and Mudalige P 2007 Mobile vehicle-to-vehicle narrow-band channel measurement and characterization of the 5.9 GHz dedicated short range communication DSRC frequency band *IEEE J. Sel. Areas Commun.* **25** 1501–16

[35] Chen A L, Wu H P, Wei Y L and Tsai H M 2016 Time variation in vehicle-to-vehicle visible light communication channels *IEEE Vehicular Networking Conf. (VNC) (Dec 2016)* pp 1–8

IOP Publishing

Visible Light Communications
Vehicular applications
Xavier Fernando and Hasan Farahneh

Chapter 3

Optical OFDM basics

VLC channels have inherent scattering and reflection properties and undergo destructive effects of multipath dispersion. This results in ISI and leads to a reduced data rate. In order to mitigate ISI, time domain equalization (TDE) technique is used in single-carrier and OFDM is used in multi-carrier communication systems [1]. OFDM has been implemented in many wireless RF and wireline standards and has also proved to be beneficial in VLC systems [2]. O-OFDM schemes have better optical power efficiency, reduced ISI and very low BER, compared to conventional optical modulation schemes such as OOK and pulse position modulation (PPM) [3].

MIMO systems have gained considerable attention in VLC networks due to their high data rate capabilities over longer distances despite multipath dispersion interference effects. The combination of OFDM and MIMO is considered a powerful physical layer solution for high-speed vehicular VLC systems, especially for accommodating bandwidth-hungry applications [4].

Generally, the combination of MIMO and OFDM has been mostly used for non-adaptive systems with a single type of modulation. Few existing works have studied V2V–VLC MIMO systems using optical OFDM. Adaptive transmission in VLC has been explored before, but mostly for indoor applications [5]. Researchers have usually assumed a Gaussian channel with fixed modulation schemes in outdoor environments.

This chapter intends to give a brief introduction on optical OFDM, from its fundamental concepts to the up-to-date experimental results. Also, we introduce a study of a novel VLC algorithm for a V2V communication system using the channel model given in section 2.3. We use O-OFDM with adaptive modulation schemes to mitigate ISI and to improve data rate. We consider DCO-OFDM and ACO-OFDM schemes for our VLC system. We consider practical reflectors; sunlight is considered as the main noise source; the effect of clipping noise is also considered. Moreover, the performance of the V2V–VLC system is optimized by singular value

doi:10.1088/978-0-7503-2284-3ch3

decomposition (SVD) technique. Also, we apply bit-loading algorithm to improve the performance of the system.

3.1 OFDM principle

Orthogonal frequency division multiplexing is a multi-carrier modulation scheme, where the individual modulated subcarriers are transmitted in parallel in a multiplexed setup. Whereas in single-carrier systems the symbol duration t_S is given by the reciprocal baud rate $1/R$, in multi carrier systems it is N_{SC}/R, where N_{SC} is the number of parallel transmitted subcarriers.

In an OFDM system, the subcarriers would overlap to achieve the highest spectral efficiency, however, they would be orthogonal to each other. This is achieved by ensuring the power of all other subcarriers is zero exactly at the frequency of the nth subcarrier. This orthogonality is achieved if the subcarrier frequencies are equidistant in frequency domain leading to the frequency of the kth subcarrier as,

$$f_k = f_0 + kf_s \tag{3.1}$$

where f_0 is the lowest frequency of the OFDM spectra, f_S is the frequency spacing, and k is the subcarriers index with $k = 0, \ldots, N_{SC}$. The superposition of the independent modulated subcarriers is typically performed by applying inverse discrete Fourier transform (IDFT), where the input channels are spaced equivalently according to (3.1). For modulation of subcarriers, phase modulation is primarily applied with four levels leading to a QPSK or higher level combined amplitude and phase modulation with QAM 8 up to QAM 64. For demodulation of the orthogonal subcarriers, a matched filter bank is required. The discrete Fourier transform (DFT) processing can replace the analog demodulator realization exhibiting desired functionality. In practice, the DFT and IDFT are realized by the fast Fourier transform (FFT) algorithm and its inverse, IFFT, exhibiting an effective realization for computation of complex values in a digital domain. Therefore, the number of sampling points in the time domain, as well as the number of grid points in the frequency domain, are a power of 2.

3.1.1 OFDM cyclic prefix

The basic concept behind the OFDM cyclic prefix is quite straightforward (figure 3.1). The cyclic prefix performs two main functions. The cyclic prefix provides a guard interval to eliminate ISI from the previous symbol. It repeats the end of the symbol so the linear convolution of a frequency-selective multipath channel can be modeled as circular convolution, which in turn may transform to the frequency domain via a discrete Fourier transform. This approach accommodates simple frequency domain processing, such as channel estimation and equalization.

The cyclic prefix is created so that each OFDM symbol is preceded by a copy of the end part of that same symbol. Different OFDM cyclic prefix lengths are available in various systems. For example, within LTE a normal length and an

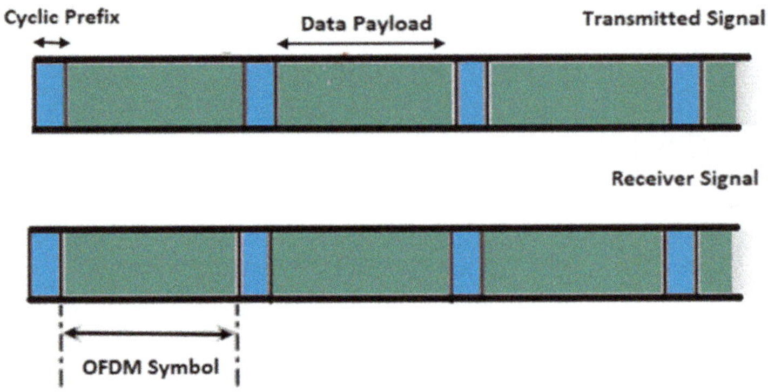

Figure 3.1. OFDM cyclic prefix.

extended length are available and after Release 8 a third extended length is also included, although not normally used.

There are several advantages and disadvantages attached to the use for the cyclic prefix within OFDM. The advantages are providing robustness and reducing ISI. However, the disadvantages of the cyclic prefix are reducing data capacity re-transmission of data. The use of a cyclic prefix is standard in OFDM and it enables the performance to be maintained even under severe multipath conditions.

3.2 Optical OFDM

In general, electrical OFDM signals are complex valued data signals. To convert a complex valued data signal into the optical domain, the signal can be electrically up-converted using an electrical intermediate frequency (IF) carrier to modulate the signal upon applying a complex electrical I/Q mixer. The resulting up-converted electrical OFDM data signal can be electro-optic (e/o) converted using a state-of-the-art amplitude modulator. Alternatively, a complex electro-optic I/Q modulator can be driven directly by the complex electrical OFDM signal and thus converted into the optical domain [6].

In optical receivers, typically, simple photodiodes are applied which operate according to the square law detection scheme. It is impossible to transfer the optical OFDM signal directly into the electrical domain. Instead of direct detection of the pure optical OFDM signal, an optical carrier must be delivered either by the transmitter (using direct detection OFDM) or by a local oscillator in the receiver (using coherent optical OFDM) via a heterodyne or intradyne approach. In the heterodyne scenario, the optical OFDM signal is converted into a real valued electrical OFDM signal at an IF. Using a subsequent electrical I/Q demodulator, the real and imaginary components are made available in the baseband. In the intradyne approach, the local oscillator wavelength is close to the transmitter wavelength. The OFDM signal then beats with the local oscillator signal in an optical 90° hybrid to obtain the I and Q components of the signal. To realize a 10 Gb/s optical OFDM system, analog-digital interfaces with a rate of 10 Giga samples per second (GSa/s) may be applied.

3.3 Related work in adaptive modulation for VLC

In literature, there has been limited research in O-OFDM–V2V–VLC systems. In this section, some of these works will be reviewed. Performance of VLC system using space-time block coding (STBC)-OFDM technique for ITS was investigated in [7]. They modeled the outdoor optical channel as a Rician channel. They proposed STBC-OFDM coding to reduce the influence of multi-path dispersion associated with the channel to achieve BER of 10^{-6} even at a low SNR. The authors in [8] presented a single-input single-output (SISO)-V2V–VLC system with DCO-OFDM. They proposed a new automotive VLC system based on the optical communication image sensor (OCI) as a receiver. Then, they applied DCO-OFDM to achieve more flexible and effective signal transmission. Also, they considered the characteristics of the LED and OCI to select the appropriate parameter values for signal processing to achieve higher data rates. The authors in [9] introduced the concept of an environment-adaptive VLC receiver for a vehicular communication system in a visible light environment to mitigate the effects of problematic conditions such as supporting long-distance communications in dynamic traffic situations and in unfriendly atmospheric conditions. In [10], the authors suggested DCO-OFDM based MIMO transmission scheme for vehicular communication. They evaluated the performances of different MIMO modes such as spatial multiplexing, and repetition code with different modulation techniques and different transmitter–receiver selection. They found the spatial multiplexing technique suffers from channel correlation, while repetition code is unsuitable for high modulation orders.

3.4 Optical OFDM scheme of a V2V–VLC system

This section outlines the investigation scheme for a V2V–VLC channel and analyzes it with various types of OFDM techniques. In this study, the model presented in section 2.3 and shown in figure 2.3 is modified to include O-OFDM with an adaptive modulation scheme. This reduces ISI and increases the data rate. Figure 3.2 shows the block diagram of the proposed optical OFDM scheme for the V2V–VLC system.

In figure 3.2, a S/P converter is used to split a binary source of data into two data streams $\mathbf{x}_1$, $\mathbf{x}_2$. Both streams are modulated using quadrature amplitude modulation (QAM) and the resulting symbols are assigned to subcarriers. Afterwards, IFFT is performed and data is converted back to a single stream using a P/S converter after adding a cyclic prefix (CP). Then, the signal is changed to unipolar using one of the O-OFDM schemes. Finally, the data is converted to analog signal $s(t)$ using a digital-to-analog (D/A) converter along with low pass filter (shaping filter).

3.4.1 DCO-OFDM modulation scheme

DCO-OFDM is a form of O-OFDM with a DC bias added. In such systems, the frequency domain signal is transformed into the time domain by using the IDFT, which can be implemented by using an IFFT. In this study, the length of the IFFT is assumed to be N. The modulated signal in the frequency domain must be conjugate

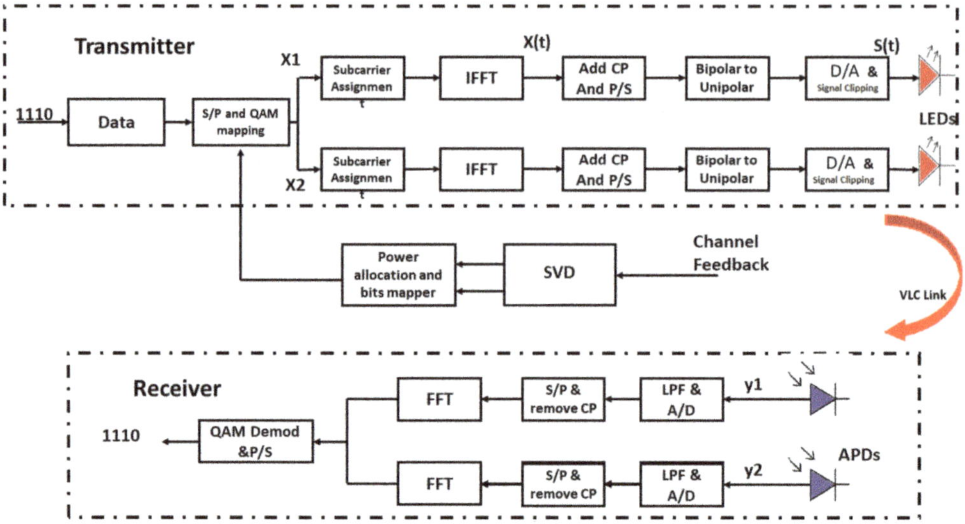

Figure 3.2. Simplified diagram of the V2V–VLC–OFDM scheme.

symmetric to ensure that the time domain signal is real. Moreover, a DC bias is added to guarantee that the transmitted signals are non-negative.

In the frequency domain, the modulated signals satisfy the following conditions:

$$\begin{cases} X(0) = X\left(\frac{N}{2}\right) = 0 \\ \qquad\qquad\qquad\qquad K = 1, 2, \ldots, N-1 \\ X(K) = X^*(N-K) \end{cases} \tag{3.2}$$

where N is the size of IFFT, and $*$ is conjugate sign.

Thus, the input vector to the IFFT block in figure 3.2 $\mathbf{x}$ can be given as

$$\mathbf{x} = \left[\mathbf{0}\ \mathbf{s}_1\ \mathbf{s}_2 \cdots \mathbf{s}_{N/2-1}\ \mathbf{0}\ \mathbf{s}^*_{N/2-1} \cdots \mathbf{s}^*_2\ \mathbf{s}^*_1\right] \tag{3.3}$$

The time domain signal can be obtained by performing IFTT and it can be expressed as

$$\begin{aligned} x[n] &= \frac{1}{N}\sum_{K=0}^{N-1} X(K)\, e^{\frac{j2\pi Kn}{N}} \\ &= \frac{1}{N}\sum_{K=0}^{\frac{N}{2}-1} X(K)\, e^{\frac{j2\pi Kn}{N}} + \frac{1}{N}\sum_{\frac{N}{2}+1}^{N-1} X(K)\, e^{\frac{j2\pi Kn}{N}} \\ &= \frac{1}{N}\sum_{K=1}^{\frac{N}{2}-1}\left(X(K)\, e^{\frac{j2\pi Kn}{N}} + X^*(K)\, e^{\frac{-j2\pi Kn}{N}}\right) \\ &= \frac{2}{N}\sum_{K=1}^{\frac{N}{2}-1}\left(a(K)\cos\left(\frac{2\pi Kn}{N}\right) - b(K)\sin\left(\frac{2\pi Kn}{N}\right)\right) \end{aligned} \tag{3.4}$$

It is assumed that $X(K) = a(K) + jb(k)$ is a zero mean complex random variable with variance σ_K^2, $X(K)$, $K = 1, \ldots, N/2 - 1$, are independent, and $a(K)$ and $b(K)$ are independent real random variables with a zero mean and variance $\sigma_K^2/2$.

According to the central limiting theorem, $x[n]$ is a Gaussian random variable with zero mean, that is $E[x(n)] = 0$, where $E[.]$ stands for expectation.

The variance of $x[n]$ can be derived as

$$\begin{aligned}\sigma_x^2 &= E[x[n]\, x^*[n]] \\ &= \frac{4}{N^2} \sum_{K=1}^{\frac{N}{2}-1} \left(\frac{\sigma_K^2}{2} \cos^2\left(\frac{2\pi Kn}{N}\right) + \frac{\sigma_K^2}{2} \sin^2\left(\frac{2\pi Kn}{N}\right)\right) \\ &= \frac{2}{N^2} \sum_{k=1}^{N/2-1} \sigma_k^2\end{aligned} \tag{3.5}$$

In DCO-OFDM technique, data is assigned to both odd and even subcarriers. A real signal can then be obtained by imposing the Hermitian symmetry property, as shown in equation (3.2). This property implies that half of the OFDM subcarriers are utilized to generate the real-time domain signal [11].

The output of D/A converter is an analog real-time signal which is shown in figure 3.3(a). This signal cannot be used to modulate the LEDs because of its bipolar nature. A certain bias value is added to this signal to convert it to unipolar and is plotted in figure 3.3(b). Afterwards, any remaining negative values are clipped at zero and are plotted in figure 3.3(c), which introduces distortion noise. The added DC value increases the transmitter power requirement, while the clipping introduces clipping noise in both the even and odd subcarriers.

The transmitted signal $s(t)$ can be expressed as

$$s(t) = x(t) + b_{\mathrm{dc}} - e_c(t) \tag{3.6}$$

where b_{dc} is the bias voltage, and $e_c(t)$ is clipping noise with a variance $\sigma_{c_{\mathrm{DCO}}}^2$.

To minimize the clipping noise, the bias voltage b_{dc} is given by [12]

$$b_{\mathrm{dc}} = 2\sigma_x\, \zeta \tag{3.7}$$

where ζ is the biasing ratio and is given by [12]

$$\zeta = \frac{b_{\mathrm{dc}}}{c_u - c_l} = \frac{-c_l}{c_u - c_l} \tag{3.8}$$

where c_u and c_l denote the upper and the lower clipping level, respectively.

The variance of the clipping noise $\sigma_{c_{\mathrm{DCO}}}^2$ is given by [13]

$$\begin{aligned}\sigma_{c_{\mathrm{DCO}}}^2 = \sigma_x^2 + \left(b_{\mathrm{dc}}^2 - \sigma_x^2\right) Q\left(\frac{b_{\mathrm{dc}}}{\sigma_x}\right) - b_{\mathrm{dc}}\sigma_x g\left(\frac{b_{\mathrm{dc}}}{\sigma_x}\right) \\ - \sigma_x^2 Q\left(\frac{-b_{\mathrm{dc}}}{\sigma_x}\right)^2 - \left(\sigma_x g\left(\frac{b_{\mathrm{dc}}}{\sigma_x}\right) - b_{\mathrm{dc}} Q\left(\frac{b_{\mathrm{dc}}}{\sigma_x}\right)\right)^2\end{aligned} \tag{3.9}$$

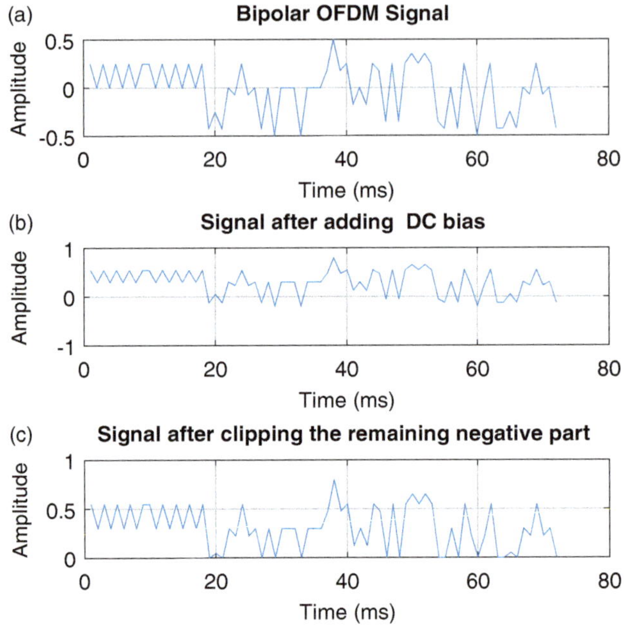

Figure 3.3. (a) OFDM bipolar signal. (b) The signal after adding DC bias voltage. (c) The signal after clipping the remaining negative.

where $Q(.)$ is the tail probability of the standard Gaussian distribution, $g(.)$ is the PDF of the clipping noise which follows Gaussian distribution and is given by [14]

$$g(u) = \frac{1}{\sqrt{2\pi\sigma_x^2}} e^{-\frac{\left(u-\frac{\zeta}{2}\sigma_x\right)^2}{2\sigma_x^2}} \tag{3.10}$$

The average transmitted optical power is expressed as

$$E[s(t)] = \sigma_x\left(\frac{\zeta}{2} + \frac{1}{\sqrt{2\pi}}e^{\frac{-\zeta^2}{8}} - \frac{\zeta}{2}Q\left(\frac{\zeta}{2}\right)\right) \tag{3.11}$$

For fixed modulation scheme, the maximum channel capacity (bps/Hz) of the DCO-OFDM–VLC system can be given as

$$R_{\text{DCO}} = \frac{1}{N + N_{\text{cp}}} \sum_{k=1}^{N/2-1} \log_2\left(1 + \frac{|H(k)|^2\,\sigma_k^2}{\sigma_n^2 + |H(k)|^2\sigma_{c_{\text{DCO}}}^2 + \sigma_{\text{ISI}}^2}\right) \tag{3.12}$$

where $|H(k)|$ indicates the kth channel transfer function, N_{cp} indicates the number of cycling prefixes of the OFDM, σ_n^2 indicates the ambient noise power which is given by (1.6), and σ_{ISI}^2 is ISI noise and is given by

$$\sigma_{ISI}^2 = \gamma^2 \left(\sum_{k=1}^{N} \int_{T_d}^{\infty} h_k(t) \otimes s(t)\, d(t) \right)^2 \quad (3.13)$$

where T_d denotes the transmitted optical pulse duration, and k is channel index.

The maximum channel capacity is subject to (optical power constraint) $0 \leqslant E[s(t)] \leqslant P_t$, where P_t is the transmitted power.

3.4.2 ACO-OFDM modulation scheme

In the ACO-OFDM technique, the modulated data is assigned to the odd indexed subcarriers, while the even indexed subcarriers are set to zero. Also, signals in this scheme can be transmitted without any DC-bias.

Referring to figure 3.2, the input vector **x** of the IFFT block can be given as

$$\mathbf{x} = \left[\mathbf{0}\ \mathbf{s}_0\ \mathbf{0}\ \mathbf{s}_1 \ldots \mathbf{s}_{N/4-1}\ \mathbf{0}\ \mathbf{s}^*_{N/4-1} \ldots \mathbf{0}\ \mathbf{s}^*_1\ \mathbf{0}\ \mathbf{s}^*_0 \right] \quad (3.14)$$

After performing IFFT, the signal can be written as

$$x\left[n + \frac{N}{2}\right] = -x[n] \quad n = 0, 1, \ldots, \frac{N}{2} - 1 \quad (3.15)$$

Also, $x[n]$ can be considered a Gaussian random variable with zero mean and variance is given by

$$\sigma_x^2 = \frac{2}{N^2} \sum_{k=1}^{N/4} \sigma_{2k-1}^2 \quad (3.16)$$

In the ACO-OFDM scheme, bipolar to unipolar conversion is achieved by clipping the negative amplitudes which will not introduce any distortion to the transmission. LED turn on voltage V_{tov}, is added to shift the unipolar signal to the dynamic range of the LED. However, any amplitude levels exceeding the upper limit of the LED dynamic range $V_{\max}$ must be clipped, thus resulting in clipping noise.

The transmitted signal in an ACO-OFDM is written as

$$s(t) = \begin{cases} x(t) & x(t) > 0 \\ 0 & x(t) \leqslant 0 \end{cases} \quad (3.17)$$

where $s(t)$ is a clipped Gaussian signal.

The PDF of $s(t)$ is written as [15]

$$f_{s(t)}(u) = \begin{cases} 0.5 & u = 0 \\ \dfrac{1}{\sqrt{2\pi\sigma_x^2}} e^{\frac{-u^2}{2\sigma_x^2}} & u > 0 \end{cases} \quad (3.18)$$

The average power of $s(t)$ can be calculated as

$$E[s(t)] = 0.5 \times 0 + \int_0^{\infty} u \frac{1}{\sqrt{2\pi\sigma_x^2}} e^{\frac{-u^2}{2\sigma_x^2}} du$$
$$= \frac{\sigma_x}{\sqrt{2\pi}} \tag{3.19}$$

The channel capacity of fixed modulation scheme for the ACO-OFDM is given by

$$R_{\text{ACO}} = \frac{1}{N + N_{cp}} \sum_{k=1}^{N/4} \log_2\left(1 + \frac{|H(k)|^2 \sigma_{2k-1}^2}{4\sigma_n^2 + |H(k)|^2 \sigma_{c_{\text{ACO}}}^2 + \sigma_{\text{ISI}}^2}\right) \tag{3.20}$$

where $\sigma_{c_{\text{ACO}}}^2$ is the variance of the clipping noise and is given by [14]

$$\sigma_{c_{\text{ACO}}}^2 = \frac{1}{2}\sigma_x^2 Q(\sqrt{2}\eta)(1 + \eta^2) - \frac{\sigma_x^2 \eta}{2\sqrt{\pi}} e^{-\eta^2} \tag{3.21}$$

where ($\eta = \frac{I_{\text{max}} - I_{\text{bais}}}{\sigma_x}$) is a dimming level, I_{max} and I_{bais} are the maximum input current and bias current of the LED, respectively.

Equation (3.20) shows that the spectral efficiency of ACO-OFDM is reduced by half as compared to DCO-OFDM, since only the odd subcarriers are used in transmission.

In this scheme, also, the maximum channel capacity is subject to (optical power constraint) $0 \leqslant E[s(t)] \leqslant P_t$.

3.5 Performance analysis and bit loading algorithm

In this section, we present a performance analysis of the V2V–VLC system using a bit loading algorithm. In this model, we use DCO-OFDM and ACO-OFDM schemes for both LOS and NLOS signal paths. APD at receiver convert optical signal to electrical signal with an addition noise. The parameter values of APD can be found in table 1.3 and in [16].

3.5.1 SNR estimation of subcarriers using SVD

The matched filter $p(t)$ is applied to the received signal $y_r(t)$ and then its output is sampled at the rate of T_s, $y_r[n] = y(nT_s)$. The discrete signal can be written as

$$y_r[n] = \sum_{s=1}^{2} (\gamma\ (s_s[n] + e_c[n]) \otimes h_{rs}[n]) + n_r[n]) \tag{3.22}$$

where s, r are the transmitter and the receiver, respectively, with values (1, 2).

The discrete signal in frequency domain can be written as

$$Y_r(k) = \sum_{s=1}^{2} (\gamma\ H_{rs}(k)\ (S_s(k) + E_c(k)) + N_r(k)) \tag{3.23}$$

The estimated channel matrix of kth subcarrier can be expressed as

$$\mathbf{H}(k) = \begin{bmatrix} H_{11}(k) & H_{12}(k) \\ H_{21}(k) & H_{22}(k) \end{bmatrix} \tag{3.24}$$

3.5.1.1 SVD-based precoding design

As RF and VLC have completely different system setups, the optimal structure of the user precoding matrix is also different. SVD based structure is optimal for RF communication systems [17], but is not optimal for VLC communication, especially when a large number of LEDs are used with a requirement of a large biasing voltage. Since we have assumed a small number of LEDs with small bias requirements in this study, SVD technique can be employed to calculate the gain of each sub-carrier. Also, this structure is practical to use in many respects; it can distribute the MIMO channel into a set of parallel non-interference channels [18]. Moreover, SVD-precoder scheme solves the channel correlation problem which causes low spatial multiplexing gain in visible light MIMO communication.

For SVD-precoder, the precoding matrix is given as

$$\mathbf{H} = \mathbf{UDV}^{*} \tag{3.25}$$

where $*$ denotes Hermitian transpose, $\mathbf{V}$ is 2×2 unitary matrix, used in the transmitter to pre-process transmitted signal, $\mathbf{U}$ is an 2×2 is the receive combining unitary matrix, and $\mathbf{D}$ is 2×2 diagonal matrix with non-negative diagonal elements (singular values of matrix $\mathbf{H}$).

Therefore, the channels are decomposed into two independent parallel subchannels. The equivalent channel gain of the two independent subchannels is given as

$$\mathbf{G}_o = \begin{bmatrix} \lambda_{11} & \lambda_{12} & \dots & \lambda_{1k} \\ \lambda_{21} & \lambda_{22} & \dots & \lambda_{2k} \end{bmatrix} \tag{3.26}$$

The gain of each subcarrier can be found from the matrix $\mathbf{G}_o$. Therefore, $SNR[k]$ of the kth subcarrier can be calculated as

$$SNR[k] = \frac{\lambda_k}{\sigma_n^2(k)} \tag{3.27}$$

By considering a targeted BER ($BER[k]$) of the kth subcarrier, the modulation order M_k can be found using (3.28) [2].

$$BER[k] = \frac{\left(\sqrt{M_k} - 1\right)}{\sqrt{M_k}\,\log_2\!\left(\sqrt{M_k}\right)}\, erfc\!\left(\sqrt{\frac{3\, SNR[k]}{2(M_k - 1)}}\right) \tag{3.28}$$

3.5.2 Adaptive transmission and bit loading scheme

Bit-loading is performed with respect to the frequency-selective channel and specific modulation order to improve the system's performance. Therefore, subcarriers having high attenuation are re-modulated with lower modulation orders. Similarly, the subcarrier having low attenuation is re-modulated with higher orders. The subcarriers of the O-OFDM signal are modulated by M – QAM; the constellation order M can differ from other subcarriers on the basis of SNR. We use table 3.2 to assign the bits for each subcarrier.

The number of bits per symbol of each subcarrier is between 1 and 8. By obtaining M_k for each subcarrier, the transmission rate (bps) for the DCO-OFDM–VLC can be calculated as

$$R_{\mathrm{DCO}_{\mathrm{Adapive}}} = \frac{1}{(N + N_{\mathrm{cp}})T_s} \sum_{k=1}^{N/2-1} \log_2 M_k \tag{3.29}$$

For the ACO–OFDM–VLC, the transmission rate can be given as

$$R_{\mathrm{ACO}_{\mathrm{Adapive}}} = \frac{1}{(N + N_{\mathrm{cp}})T_s} \sum_{k=1}^{N/4-1} \log_2 M_k \tag{3.30}$$

3.6 Results and discussions

3.6.1 Simulation environment

In this study, MATLAB software is used to model the V2V–VLC system, and parameters/values used in simulations are summarized in tables 1.3, 2.1, and 3.1. Multi-path propagation with one reflection system is considered assuming that all the reflectors surfaces have the same reflective factor. Moreover, summer season with clear sky is considered. Also, vehicles are moving at the same speed in the same

Table 3.1. Simulation parameters.

The parameter	Value
Toronto location	43.6532°N, 79.3832°W
Weather	Clear sky
Number of FFT points	64
Subcarrier modulation scheme	4-QAM, 16-QAM,...,256-QAM
Signal bandwidth	50 MHz
Cyclic prefix length CP	8
Targeted BER	10^{-5}
Biased voltage (V)	8 dB, 10 dB
Vehicles speed (same direction)	30 km/h

Table 3.2. Look-up-table.

	Number of bits/subcarrier	
M_k	DCO-OFDM–SNR (dB)	ACO-OFDM–SNR (dB)
0	< 7	< 3
1	7	3
2	8	4
3	9	5
4	10	6
5	11	7
6	12	8
7	> = 14	10
8	NA	> = 12

direction. The city of Toronto is considered for sunlight intensity calculations, whose location coordinates are given in table 3.1. We do Monte Carlo simulation for 5000 iterations for simulating the V2V–VLC channel [19]. To reduce the impact of the channel causing signal distortion, a truncated sync pulse filter is used for pulse shaping. Also, We use a Philips Ultinon LED 12985BWX2 as transmitter [20]. Moreover, the Hamamatsu-S8664 APD is considered as a receiver [16]. The scheme proposed in [21] is used to model the nonlinearity of the LED's *I–V* property. Figure 3.4 shows that the simulated *I-V* property fits well with the data obtained from the LED datasheet. Moreover, the linear *I–V* curve is obtained by linear regression of the data.

3.6.2 Results and analysis

The system performance is analyzed by considering various parameters for both types of the O-OFDM scheme. The distributions of bit- and power-loading of the DCO-OFDM for 64 subcarriers at luminance intensity (1000–2500 (cd)) with an 8 dB and 10 dB bias is shown in figure 3.5. A 128 QAM (7 bits/symbol) is assigned to subcarriers having an SNR $\geqslant$ 14 dB, binary phase shift keying (BPSK) (1 bit/symbol) is chosen for the subcarriers having SNR = 7 dB (as seen with 8 dB bias) and no bits assigned for SNR < 7 dB. The transmitter can obtain the modulation order M_k to be employed for each subcarrier by using table 3.2. Also, figure 3.5 shows clipping noise is decreased with the increase in bias. Therefore, for 10 dB bias, as the clipping noise is less, the total noise power will depend more on channel frequency response than on the 8 dB bias. As a result of the 8 dB bias, the bit allocation decreases by two bits over 64 subcarriers. Moreover, figure 3.5 shows the power distribution and modulation order of the subcarriers has decreasing trends at a high frequency spectrum. The reason for this type of behavior depends on the low-pass property of the LED. In figure 3.6, bit loading and power distribution for different subcarriers

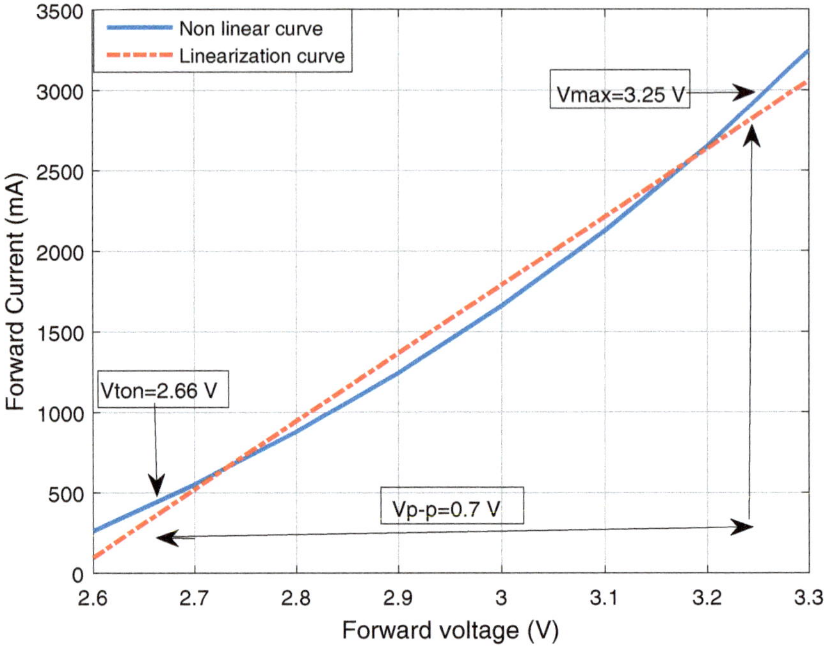

Figure 3.4. *I–V* characteristic curve of the LED.

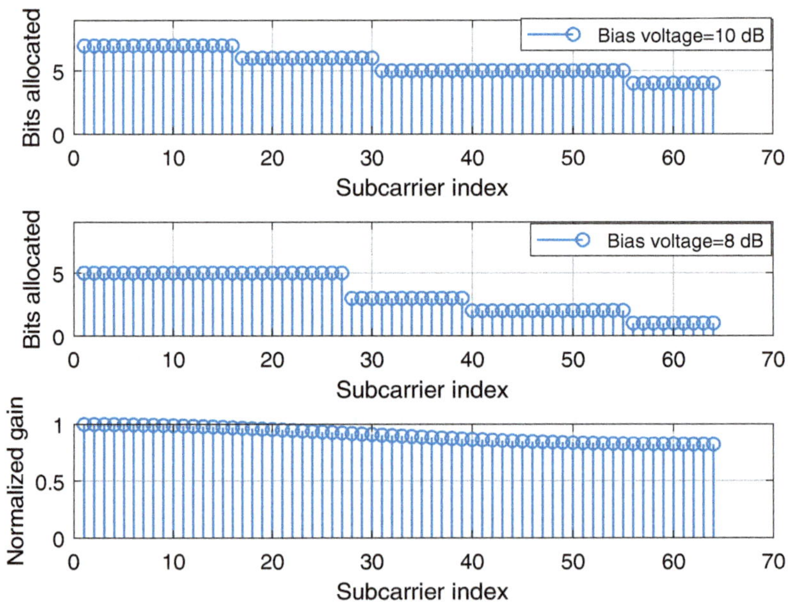

Figure 3.5. Number of loaded bits for DCO-OFDM when $I_T(\alpha_T, \beta_T) = 1500$ (cd).

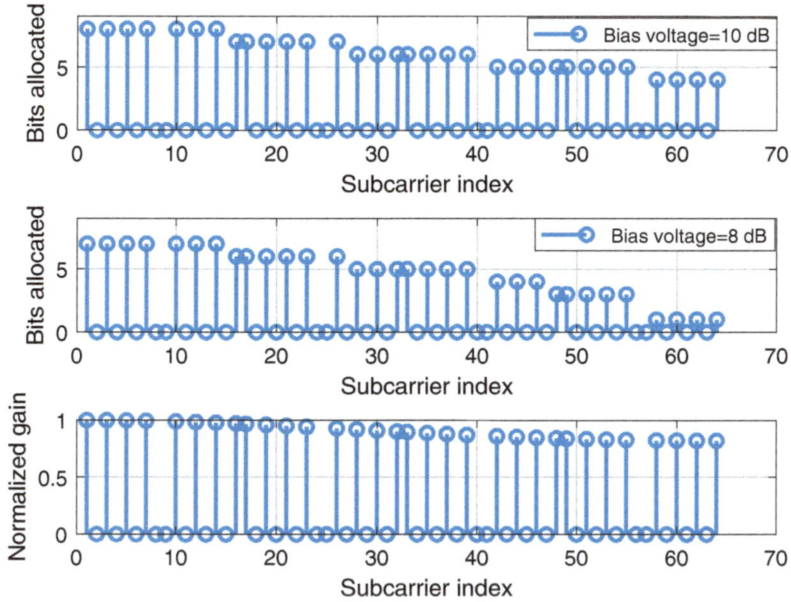

Figure 3.6. Number of loaded bits for ACO-OFDM when $I_T(\alpha_T, \beta_T) = 1500$ (cd).

for the ACO- OFDM adaptive modulation scheme are presented, under similar channel conditions as in the DCO-OFDM with an 8 dB bias and 10 dB bias, respectively. In this technique, only the odd subcarriers are used for data transmission. Also, the power distribution and modulation order of the subcarriers are decreasing with the high frequency spectrum. Moreover, a maximum of 256 QAM (8 bits/symbol) is assigned to subcarriers having an SNR of about 12 dB, whereas the BPSK (1 bit/symbol) format is chosen for the subcarriers having SNR = 3 dB, and no bits assigned for SNR < 3 dB. Note that, the number of assigned bits to the ACO-OFDM scheme is higher than for the DCO-OFDM scheme. This is because the DCO-OFDM experiences higher clipping distortion compared to ACO-OFDM at certain illuminance intensity. Therefore, the number of loaded bits per subcarrier for the DCO-OFDM is generally lower than in the ACO-OFDM. Also, figure 3.6 shows that increasing the bias decreases the clipping noise. As a result of an 8 dB bias, the bit allocation decreases by one bit over 32 subcarriers [19].

3.6.2.1 Transmission rate analysis

In figures 3.7 and 3.8, transmission rates for adaptive DCO-OFDM and ACO-OFDM, are presented. The adaptive transmissions are compared with the fixed modulation schemes while maintaining the desired BER of 10^{-5}. In figure 3.7, it is shown that by employing adaptive transmission in the DCO-OFDM, data rate is significantly improved, compared to fixed modulation. Furthermore, the clipping distortion affects the performance of the DCO-OFDM, above luminance intensity of 1600 (cd). In figure 3.8, the transmission rate for the ACO-OFDM is calculated

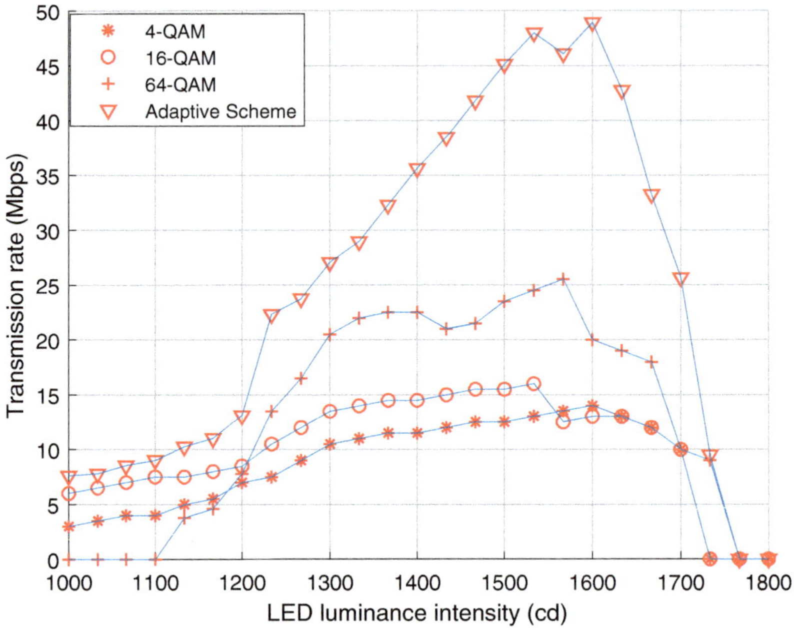

Figure 3.7. The transmission rate of DCO-OFDM with adaptive modulation.

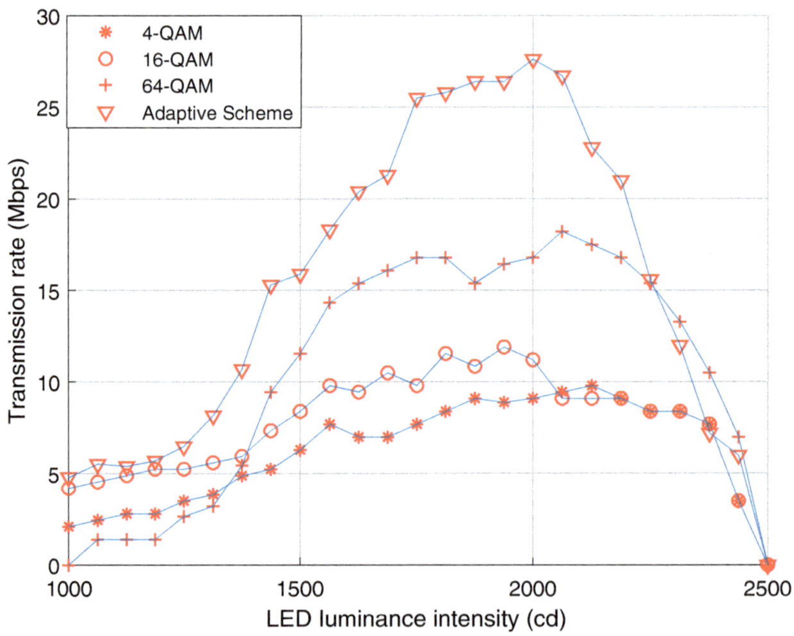

Figure 3.8. The transmission rate of ACO-OFDM with adaptive modulation.

and it is observed that the clipping distortion degrades the performance when the luminance intensity exceeds 2000 (cd). Moreover, the maximum achievable rate for the ACO-OFDM is lower than the DCO-OFDM for both types of transmission scheme, i.e. adaptive and fixed. The ACO-OFDM utilizes a dynamic range more effectively due to half the number of data subcarriers [19].

3.6.2.2 BER analysis

BER performance for both, the ACO-OFDM and the DCO-OFDM is investigated, for adaptive and fixed modulation schemes. Figures 3.9 and 3.10 show better SNR performance for adaptive modulation than fixed modulation, for both schemes. Moreover, it can be seen that adaptive modulation schemes can meet the target of BER at less SNR value than other QAM schemes. Also, the BER performance is not recorded for SNR range of (0–3 dB) for the ACO-OFDM, and for SNR range of (0–7 dB) for the DCO-OFDM because no bits are transmitted at all in these ranges (according to the bit loading algorithm). Moreover, the BER performance is not recorded for SNR > 18 dB for both schemes because the clipping distortion degrades the performance. However, figures 3.9 and 3.10 show better performance of the ACO-OFDM compared with the DCO-OFDM. This happens due to the inherent characteristics of the DCO-OFDM, being sensitive to clipping noise [19].

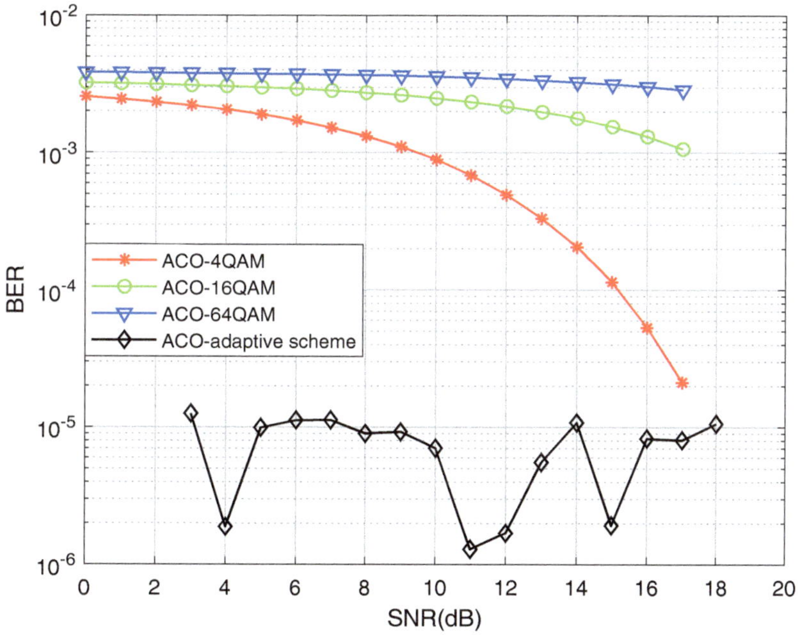

Figure 3.9. SNR versus BER for ACO-OFDM with FFT size = 64.

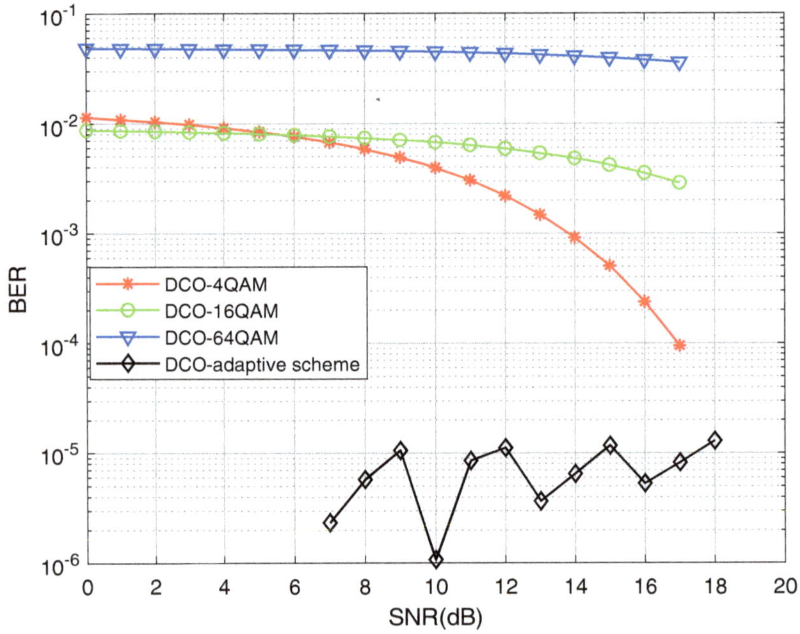

Figure 3.10. SNR versus BER for DCO-OFDM with FFT size = 64.

3.7 Chapter summary

This chapter investigates and analyzes adaptive modulation schemes for the O-OFDM based V2V–VLC system considering the practical noise environment to fully exploit the transmission resources. Two modulation schemes are used, DCO-OFDM and ACO-OFDM to combat ISI and to improve the data rate. We compare the performance of the adaptive modulation schemes with fixed modulation schemes in terms of BER, and data rate, where the adaptive schemes show superiority in performance.

References

[1] Uysal M, Ghassemlooy Z, Bekkali A, Kadri A and Menouar H 2015 Visible light communication for vehicular networking: Performance study of a V2V system using a measured headlamp beam pattern model *IEEE Veh. Technol. Mag.* **10** 45–53

[2] Elgala H, Mesleh R and Haas H 2011 On the performance of different OFDM based optical wireless communication systems *IEEE/OSA J. Opt. Commun. Net.* **3** 620–8

[3] Zhang X, Zhou Y F, Yu Y P, Han P C and Wang X R 2014 Comparison and analysis of DCO-OFDM, ACO-OFDM and ADO-OFDM in IM/DD systems *Appl. Mech. Mater.* **12** 701–2

[4] Wang Q, Wang Z and Dai L 2015 Multiuser MIMO-OFDM for visible light communications *IEEE Photonics J.* **7** 1–11

[5] Dang J, Wu L, Zhang Z and Liu H 2015 Adaptive modulation schemes for visible light communications *J. Lightwave Technol.*

[6] Dischler R, Buchali F and Liu X 2009 Optical ofdm: A promising high-speed optical transport technology *Bell Labs Tech. J.* **14** 125–46

[7] Lee K, Li C, Yi Y and Lee K 2014 Performance analysis of visible light communication using the STBC-OFDM technique for intelligent transportation systems *Int. J. Electron.* **101** 1117–33

[8] Goto Y, Takai I, Yamazato T, Okada H, Fujii T, Kawahito S, Arai S, Yendo T and Kamakura K 2016 A new automotive VLC system using optical communication image sensor *IEEE Photonics J.* **8** 1–17

[9] Cailean A-M and Dimian M 2016 Toward environmental-adaptive visible light communications receivers for automotive applications: A review *IEEE Sens. J.* **16**

[10] Turan B, Narmanlioglu O, Ergen S C and Uysal M 2016 Broadcasting brake lights with MIMO-OFDM based vehicular VLC *IEEE Veh. Net. Conf.* 1–7

[11] Hong Y, Viterbo E and Fernando N 2012 Flip-OFDM for unipolar communication systems *IEEE Trans. Commun.* **60** 3726–33

[12] Zhou G T, Yu Z and Baxley R J 2012 EVM and achievable data rate analysis of clipped OFDM signals in visible light communication *EURASIP J. Wirel Commun. Net.*

[13] Zhao C, Ling X, Wang J, Liang X and Ding Z 2016 Offset and power optimization for DCO-OFDM in visible light communication systems *IEEE Trans. Signal Process.* **64** 349–63

[14] Lu Q f, Ji X s and Huang K z 2014 Clipping distortion analysis and optimal power allocation for ACO-OFDM based visible light communication *4th IEEE Int. Conf. Information Sci. Technol. (April 2014)* pp 320–3

[15] Dang J, Wu L, Zhang Z and Liu H 2015 Adaptive modulation schemes for visible light communications *J. Lightwave Technol.* **33** 117–25

[16] Photo diode parameters http://www.hamamatsu.com

[17] Palomar D P, Cioffi J M and Lagunas M A 2003 Joint Tx-Rx beamforming design for multicarrier MIMO channels: a unified framework for convex optimization *IEEE Trans. Signal Process.* **51** 2381–401

[18] Wang R, Gao Q, You J, Liu E, Wang P, Xu Z and Hua Y 2017 Linear transceiver designs for MIMO indoor visible light communications under lighting constraints *IEEE Trans. Commun.* **65** 2494–508

[19] Farahneh H, Hussain F and Fernando X 2018 Performance analysis of adaptive ofdm modulation scheme in vlc vehicular communication network in realistic noise environment *EURASIP J. Wirel. Commun. Net.* **2018** 243

[20] Light emiting diode http://www.powerbulbs.com

[21] Elgala H, Mesleh R and H Haas 2010 An LED model for intensity-modulated optical communication systems *IEEE Photonics Technol. Lett.* **22** 835–7

IOP Publishing

Visible Light Communications
Vehicular applications
Xavier Fernando and Hasan Farahneh

Chapter 4

Precoder and equalizer in 2 × 2 MIMO VLC systems

Precoding is a technique which exploits transmit diversity by weighting the information stream, i.e. the transmitter sends the coded information to the receiver to achieve pre-knowledge of the channel. VLC channels inherently exhibit frequency-selective and multi-path fading effects that result in ISI and thus reduce the data rate. To this end, optical MIMO can provide spatial diversity and thus achieve a high data rate. Inspired by these facts, we present different precoding and equalization schemes for 2 × 2 MIMO for the V2V–VLC system. In this study, we consider three detection schemes, ZF, ML, and MMSE. Also, we consider flickering/dimming control and the nature of outdoor VLC channel as a frequency selective channel. Moreover, we investigate a transceiver considering perfect channel state information (CSI) based on different detection algorithms.

4.1 Related work in precoding and equalization for VLC

There is limited published work in precoding and equalization areas in the MIMO–V2V–VLC System. Most of the existing work on the precoding and equalization of VLC systems is mainly limited to indoor environments. Tomas *et al* [1] designed a transceiver model for realistic simulation of vehicular VLC. They tested the design of a VLC transceiver prototype using a modified version of the 802.11 MAC protocol, showing that a MAC protocol specifically designed for VLC should take into account its specific physical properties. In [2], the authors studied a linear transceiver design for indoor VLC with multiple LEDs; they showed that for MIMO-VLC, the optimal precoding reduces to a simple LED selection strategy. Corresponding channel structure and simple receiver design for VLC systems were also discussed in [3]. The MIMO transceiver was optimized to improve the system performance in [4]. The power and the positive offset are jointly designed to improve the spectral efficiency by taking BER requirement, non-negativity constraint and

doi:10.1088/978-0-7503-2284-3ch4

summing optical power constraint of the transmitted signals. The authors in [5] investigated a joint precoding matrix and receiving matrix design via a convergence guaranteed iterative algorithm by taking the positive constraint on the transmitted signals into account. In another work [5], Ying *et al* investigated the design of optical wireless MIMO precoder and equalizer systems for indoor VLC. The authors studied the system with perfect channel state and imperfect channel state information. In [6], a multi-user downlink channel was considered and the corresponding precoding design was optimized. The authors imposed the ZF structure on the precoding matrix. In [7], power and offset allocation with adaptive modulation were well discussed in SVD-based MIMO systems for optical wireless channels, but a dimming control was not supported. Authors in [8] proposed a modified SVD-VLC MIMO system to meet the illumination constraints, but indicated that additional adjusting should be applied to the input signals so that both the non-negativity of the intensity modulation and dimming control could be satisfied. In the area of the MIMO VLC system, the precoder and equalizer design are still open to researchers, especially in outdoor-VLC applications.

4.2 Dimming control technique and its performance in VLC systems

Illumination and communication are the two main functions of the LED lamps in VLC systems. The brightness of the LED light needs to be adjusted in accordance with the requirements and comfort of users and to save energy. The main dimming control technique is pulse width modulation (PWM) where the brightness of the LED light is changed by adjusting the duty cycle of the PWM signal without varying the LED current [9]. As shown in figure 4.2, the LED current is modulated by a PWM signal to control its brightness by changing the on duration within the whole period. So the light is dimming during the whole period of the PWM signal. The data are modulated onto the dimming controlled light in the on time only, and no light is transmitted during the off time. Since the LED current remains constant throughout, the brightness of the LED light is varied by applying the PWM dimming control signal to adjust the duration of its on period as a fraction of the whole period. With the duty cycle of the PWM signal set to 1, all the LED light is transmitted and the light obtained is of the highest brightness. When the duty cycle is reduced, the LED light in the off period is blocked and hence the light is dimming during the whole duration of the PWM signal. It is worth mentioning that the frequency of the PWM signal should be higher than 200 Hz, otherwise it will cause flicker and would have adverse effects on user health [10, 11].

4.3 Precoder and equalizer in 2 × 2 MIMO V2V–VLC systems

Considering the system given in section 2.3 and shown in figure 2.3, the precoder–equalizer system for the V2V–VLC system is shown in figure 4.1, where the binary data modulated, then precoded by the precoder **G**, then DC-biased is added to ensure that the transmitted signal is positive. To this end, the signal is transmitted through an optical channel. In the receiver, the received signal is equalized by equalizer **E**, then passes to be demodulated to predict the transmitted data.

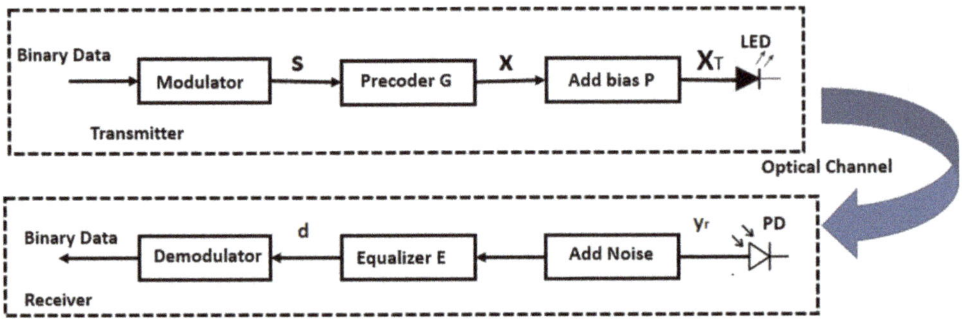

Figure 4.1. Illustration of the investigated precoder and equalizer system.

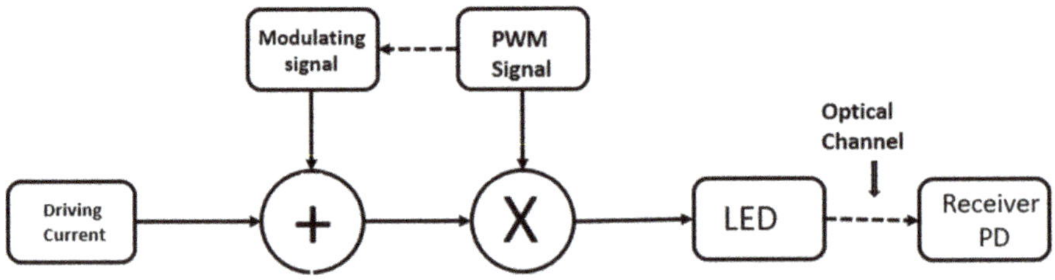

Figure 4.2. Dimming control.

4.3.1 Investigating the transmitter

Since we have 2 × 2 MIMO systems, the original information bits are modulated into source data vector denoted by [12]

$$\mathbf{s}(t) = [s_1(t)s_2(t)]^T \tag{4.1}$$

where any bit s_i is assumed to be return to zero (RZ) with zero mean and two levels $[-A, A]$. The source data vector will be multiplied by a 2 × 2 precoding matrix **G**, where **G** is given as

$$\mathbf{G} = \begin{bmatrix} g_{11} & g_{12} \\ g_{21} & g_{22} \end{bmatrix} \tag{4.2}$$

Each entry of **G** denotes the precoding weight of the transmitted signal between the transmitter s and the receiver r, $(r, s = 1, 2)$.

The precoded data vector **x** is given by

$$\mathbf{x} = \mathbf{G}\mathbf{s} \tag{4.3}$$

It is worth noting that since VLC is based on intensity modulation, the vector **x** should be positive. Thus, a DC biasing vector $\mathbf{p} = [p_1 p_2]^{\mathrm{T}}$ is added to the transmitted signal to ensure that the transmitted signal is positive, which guarantees that each transmitted vector $(\mathbf{x}_{\mathrm{T}_1}, \mathbf{x}_{\mathrm{T}_2})$ is positive.

$$\mathbf{x}_{\mathrm{T}} = \mathbf{G}\mathbf{s} + \mathbf{p} \tag{4.4}$$

The transmitted data vector $\mathbf{x_T}$ is calculated for the VLC system as follows

$$\begin{bmatrix} x_{T_1} \\ x_{T_2} \end{bmatrix} = \begin{bmatrix} g_{11} & g_{12} \\ g_{21} & g_{22} \end{bmatrix} \begin{bmatrix} s_1 \\ s_2 \end{bmatrix} + \begin{bmatrix} p_1 \\ p_2 \end{bmatrix} \tag{4.5}$$

Let l and u be the lower and the upper levels of the LED dynamic range, respectively. For each LED, we assume that $[l, u](0 < l)$. It is the dynamic-range constraint, and in order to meet the dimming control, i.e., to achieve the illumination requirement by controlling the output power, the following condition must be fulfilled, $l \leqslant x_{T_1}, x_{T_2} \leqslant u$.

Thus, the mean value of any transmitted vector is given by

$$E[x_{T_s}] = \sum_{s=1}^{2} \sum_{r=1}^{2} g_{rs}\, E[s_s] + E[p_s] = p_s \tag{4.6}$$

Equation (4.6) shows that the transmitted signal average is positive and it thus ensures that the transmitted signal is positive, $E[s_s] = 0$ is the mean value of the transmitted bit, $E[p_s]$ is the mean value of the added bias voltage, and s indicates the transmitter and r indicates the receiver with values 1 or 2.

To keep the transmitted signal within the LED dynamic range, the following two constraints should be satisfied $\sum_{s=1}^{2}\sum_{r=1}^{2}|g_{rs}|\, A \leqslant u - p$, and $\sum_{s=1}^{2}\sum_{r=1}^{2}|g_{rs}|\, A \leqslant p - l$.

To meet the brightness control of the vehicle's headlights

$$abs(\mathbf{G})A \leqslant min(u - p, p - l) \tag{4.7}$$

The DC bias vector can affect the performance of precoder and equalizer. The BER performance is best when the DC bias is the midpoint of l and u. That is, $p = \frac{1}{2}(l + u)$, we can get a high-performance precoder and equalizer under the constraint of (4.7) [5].

4.3.2 Investigating the receiver

At the receiver, PDs generate an electric signal proportional to the intensity of the received optical signals. Then, the received signal vector can be expressed as follows [12]

$$\mathbf{y_r} = \gamma \mathbf{H} \mathbf{x_T} + \mathbf{n} \tag{4.8}$$

where $\mathbf{y_r}$ is the received vectors, $\mathbf{H}$ indicates the channel matrix comprises of LOS and NLOS components and $\mathbf{n}$ is the total noise vector. Here, we consider an AWGN following the distribution of $N(0, \sigma^2 I)$.

For 2 × 2 MIMO, the channel matrix $\mathbf{H}$ is given by

$$\mathbf{H} = \begin{bmatrix} h_{11} & h_{12} \\ h_{21} & h_{22} \end{bmatrix} \tag{4.9}$$

Since we have LOS path and NLOS path coming by the reflectors within the test area, each entry of matrix **H** can be given as

$$h(t) = h_{\mathrm{LOS}}(t) + h_{\mathrm{NLOS}}(t) \tag{4.10}$$

Considering a two-dimensional (2-D) system, i.e all sources and receivers lie in the same plane, each entry of the LOS path in **H** is given by

$$h_{rs} = \frac{\cos(\phi_{rs})\cos(\theta_{rs})\, A_r T_s(\theta_{rs})\, g(\theta_{rs})}{\pi R_{\mathrm{rs}}^2}. \tag{4.11}$$

where ϕ_{rs} denotes the emitting angle between source s and receiver r, θ_{rs} denotes the incident angle between source s and receiver r, A_r denotes the effective area of the receiver, $T_s(\theta_{\mathrm{rs}})$ denotes the signal transmission coefficient of an optical filter, and $g(\theta_{\mathrm{rs}})$ denotes the concentrator gain.

Any entry in the channel matrix **H** for NLOS path between transmitter s and receiver r through reflector q can be calculated as follows

$$h_{\mathrm{rs(NLOS)}} = h_{qs} + h_{rq} \tag{4.12}$$

It is to be noted that the NLOS gain factor denoted by $h_{\mathrm{rs(NLOS)}}$ consists of two components, the gain between the transmitter s and reflector q, denoted by h_{qs}, and the gain between the reflector q and the receiver r, denoted by h_{rq}. These gains are calculated as follows

$$h_{qs} = \frac{\cos(\phi_{qs})\cos(\theta_{qs})\, A_q}{\pi R_{qs}^2} \tag{4.13}$$

where ϕ_{qs} denotes the emitting angle between source s and reflector surface q, θ_{qs} denotes the incident angle between source s and reflector surface q, R_{qs} denotes the distance between source s and reflector surface q, A_q denotes the effective area of the reflector surface q.

The gain factor h_{rq} is given by

$$h_{rq} = \rho\,\frac{\cos(\phi_{rq})\cos(\theta_{rq})\, A_r T_r(\theta_{rq})\, g(\theta_{rq})}{\pi R_{rq}^2}. \tag{4.14}$$

where ϕ_{rq} is the emitting angle between reflector q and receiver r, θ_{rq} is the incident angle between receiver r and reflector surface q, R_{rq} denotes the distance between receiver r and reflector surface q, and ρ is the reflectivity of the reflector surface.

Equation (4.8) can be expanded to two received signals, one from the LOS, and another one from NLOS, $\mathbf{y_r} = \mathbf{y_{r_{LOS}}} + \mathbf{y_{r_{NLOS}}}$, where $\mathbf{H} = \mathbf{H}_{\mathrm{LOS}} + \mathbf{H}_{\mathrm{NLOS}}$.

The received signal, which is given by (4.8), can be written as

$$\mathbf{y_r} = \gamma\mathbf{H}(\mathbf{Gs} + \mathbf{p}) + \mathbf{n} \tag{4.15}$$

At the receiver side, the term **Hp** is subtracted from $\mathbf{y_r}$ before the equalization. After this subtraction, the communication model can be written as

$$\hat{\mathbf{y}}_\mathbf{r} = \gamma\mathbf{HGs} + \mathbf{n} \tag{4.16}$$

The estimated symbols by the equalizer can be expressed as follows

$$\begin{aligned}\hat{\mathbf{s}} &= \mathbf{E}(\mathbf{y}_r - \mathbf{HP}) \\ &= \mathbf{E}(\gamma\mathbf{HGs} + \mathbf{n})\end{aligned} \tag{4.17}$$

4.3.3 Equalization matrix design

An equalizer is a digital filter that is used to mitigate the effects of intersymbol interference that is introduced by a time-dispersive channel. In order to recover the source data from the received signal, we can formulate the problem as follows: 'design a precoder matrix **G** at the transmitter and an equalizer matrix **E** at the receiver to minimize the MSE between the transmitted data and the recovered data', that is,

$$\begin{aligned}&\underset{\mathbf{G},\mathbf{E}}{\text{MSE}} \quad (\mathbf{d}, \mathbf{s}, \mathbf{G}, \mathbf{E}) \\ &\text{subject to } abs(\mathbf{G})\, A \leqslant min(u - p, p - l)\end{aligned}$$

where $\mathbf{d} = \gamma\mathbf{EHGs} + \mathbf{En}$.

Here, we assume that the channel matrix **H** is perfectly known. Also, we assume the data vector $\mathbf{s}_s$ is taken from one of an M-pulse amplitude modulation (PAM) symbol with $M = 2^k$ and k being the number of bits per symbol.

4.3.3.1 Zero forcing equalizer (ZF)

The ZF equalizer is a linear equalization algorithm used in communication systems, which inverts the frequency response of the channel. This algorithm applies the inverse of the channel to the received signal, to restore the signal before the channel. In this method, a linear time-invariant filter is used and the ISI component at the output of the equalizer is forced to zero. This will be useful when ISI is significant compared to noise.

The ZF equalization matrix is given by [13]

$$\mathbf{E}_{ZF} = (\mathbf{H}^H\mathbf{H})^{-1}\,\mathbf{H}^H \tag{4.18}$$

The $\mathbf{E}_{ZF}$ matrix should have non-zero off diagonal elements and hence this cancels out the interference signal. In fact, it is simple to implement and easy to analyze. The main disadvantage of this equalizer is amplifying the noise. Using ZF equalization, the estimated transmitted symbol vector is [13]

$$\begin{aligned}\hat{\mathbf{s}} &= \gamma\mathbf{E}_{ZF}\,\mathbf{y}_r \\ &= \gamma(\mathbf{H}^\dagger\mathbf{HGs} + \mathbf{H}^\dagger\mathbf{n}) \\ &= \gamma(\mathbf{Gs} + \mathbf{H}^\dagger\mathbf{n})\end{aligned} \tag{4.19}$$

where $(.)^\dagger$ means the pseudo inverse of a matrix.

4.3.3.2 Maximum-likelihood equalizer (ML)

ML detection calculates the Euclidean distance between the received signal vector and the product of all possible transmitted signal vectors with the given channel **H**, and finds the one with the minimum distance. ML detection determines the estimate of the transmitted signal vector. The maximum of the likelihood function is achieved when $|\mathbf{y}_r - \mathbf{H}\hat{\mathbf{s}}|^2$ reaches the minimum value, therefore the estimation of the ML detector is given by [14]

$$\mathbf{s}_{\hat{\mathrm{ML}}} = \mathbf{arg} \min_{\mathbf{s}_{\hat{\mathrm{ML}}} \in \mathbf{S}^{\mathrm{M}}} \|\mathbf{y}_r - \mathbf{H}\hat{\mathbf{s}}\|^2 \tag{4.20}$$

4.3.3.3 Minimum mean squared error equalizer (MMSE)

For MMSE equalizer we have [15]

$$\begin{aligned} \mathrm{MSE}(\mathbf{d}, \mathbf{s}, \mathbf{G}, \mathbf{E}) &= E[\|\mathbf{d} - \mathbf{s}\|^2] \\ &= E[\|(\mathbf{EHG} - \mathbf{I})\,\mathbf{s} + \mathbf{En}\|^2] \\ &= Tr((\mathbf{EHG} - \mathbf{I})\mathbf{D}(\mathbf{EHG} - \mathbf{I})^{\mathbf{H}}) + \mathbf{Tr}(\mathbf{ER}_n\mathbf{E}^{\mathbf{H}}) \\ &= Tr((\mathbf{EHGD})(\mathbf{EHG})^{\mathbf{H}}) + \mathbf{Tr}(\mathbf{ER}_n\mathbf{E}^{\mathbf{H}}) + \mathbf{Tr}(\mathbf{D}) \\ &\quad - \mathbf{Tr}(\mathbf{EHGD}) - \mathbf{Tr}(\mathbf{D}(\mathbf{EHG})^{\mathbf{H}}) \end{aligned} \tag{4.21}$$

where $\mathbf{D} = E[\mathbf{s}\,\mathbf{s}^{\mathbf{H}}]$, $\mathbf{R}_n = \mathbf{E}[\mathbf{nn}^{\mathbf{H}}] = \boldsymbol{\sigma}_n^2\mathbf{I}_2$, and $\mathbf{D} = \mathrm{diag}[D, D]$, where D is given by

$$D = \frac{A^2(M + 1)}{3(M - 1)} \tag{4.22}$$

Here, we assume the data vector $\mathbf{s}_s$ is taken from one M-PAM symbol with $M = 2^k$ and k being the number of bits per symbol.

The optimal $\mathbf{E}$ for given $\mathbf{G}$ satisfies the following condition

$$\frac{dMSE(\mathbf{d}, \mathbf{s}, \mathbf{G}, \mathbf{E})}{d\mathbf{E}} = 0 \tag{4.23}$$

Then the MMSE equalizer is given by

$$\mathbf{E} = \mathbf{DG}^{\mathbf{H}}\mathbf{H}^{\mathbf{H}}\left(\mathbf{HGDG}^{\mathbf{H}}\mathbf{H}^{\mathbf{H}} + \boldsymbol{\sigma}_n^2\mathbf{I}_2\right)^{-1} \tag{4.24}$$

The related MSE covariance matrix can be given as

$$\begin{aligned} \mathfrak{R} &= E[(\mathbf{s} - \hat{\mathbf{s}})(\mathbf{s} - \hat{\mathbf{s}})^{\mathbf{T}}] \\ &= \left(\mathbf{D}^{-1} + \frac{\mathbf{1}}{\boldsymbol{\sigma}_n^2}\mathbf{G}^{\mathbf{H}}\mathbf{H}^{\mathbf{H}}\mathbf{HG}\right)^{-1} \end{aligned} \tag{4.25}$$

4.3.4 Precoding matrix

4.3.4.1 MMSE precoder

Since the MMSE equalization technique is investigated, here we will use the corresponding precoder matrix which is given by [16]

$$\mathbf{G}_{\mathrm{MMSE}} = \sqrt{\frac{1}{2}}\mathbf{V}\triangle\tilde{\mathbf{D}} \tag{4.26}$$

where $\mathbf{V}$ comes from the decomposition of channel matrix $\mathbf{H}^*\Phi_{\mathbf{nn}}^{-1}\mathbf{H}$ and $\mathbf{H} = \mathbf{V\Theta V}^*$, $\mathbf{\Phi_{nn}} = \mathbf{2\sigma_n^2 I}$ is the noise matrix, $\tilde{\mathbf{D}}$ is a normalized discrete Fourier transform matrix, and $\mathbf{\Delta}$ is a diagonal matrix with the diagonal elements determined by [17]

$$\|m_{ii}\| = \frac{1}{2}\left(1 + \sum_{j=1}^{2}\theta_j^{-1}\right) - \theta_i^{-1} \tag{4.27}$$

where θ_i are diagonal elements of Θ coming from the eigenvalue decomposition $\mathbf{H}^* \Phi_{\mathbf{nn}}^{-1}\mathbf{H} = \mathbf{V\Theta V}^*$.

4.3.4.2 ZF precoder
The precoder for ZF equalization is [13]

$$\mathbf{G_{ZF}} = \sqrt{\frac{\mathbf{1}}{\mathbf{2tr}(\mathbf{\Theta}^{-1/2})}}\mathbf{V\Theta}^{-1/4}\tilde{\mathbf{D}} \tag{4.28}$$

4.3.4.3 ML precoder
The precoder for ML equalization is given by [14]

$$\mathbf{G_{ML}} = \begin{bmatrix} g_{m11} & g_{m12} \\ g_{m21} & g_{m22} \end{bmatrix} \tag{4.29}$$

subject to the following constraints $g_{m11}\, g_{m21} + g_{m12}\, g_{m22} = 0$ and $g_{m11}^2 + g_{m12}^2 + g_{m21}^2 + g_{m22}^2 = 0$.

4.4 Simulation and results

In this study, we use MATLAB to model the system and simulation parameters are summarized in tables 1.3 and 2.1. In the simulations, we consider multi-path propagation with one reflection system where all reflectors' surfaces are assumed to have the same reflective factor. Moreover, without loss of generality, we consider a clear sky and the vehicles are moving at uniform speed in the same direction. We use a Philips Ultinon LED 12985BWX2 as transmitter [18]. Also, we use the APD model (Si) Hamamatsu $S8664 - 1010$ as a receiver [19]. Moreover, we also consider 4-PAM modulated symbols and the constellation of 4-PAM is formed in the range of $[-3, 3]$ (i.e. $A = 3$).

The investigated optimal precoder and equalizer can be easily applied to the practical design. We generate and fix the NLOS gain as given by (2.40). Both the signals and the noises are assumed to be independent, which is also reasonable in V2V–VLC practical systems. Additionally, we consider the ambient noise model used in [20]. To measure the performance of the different schemes, we compare the performance of the investigated precoders and equalizers in terms of BER and MSE. Notice that the power constraint in (4.7) is still valid for all the obtained simulations [12].

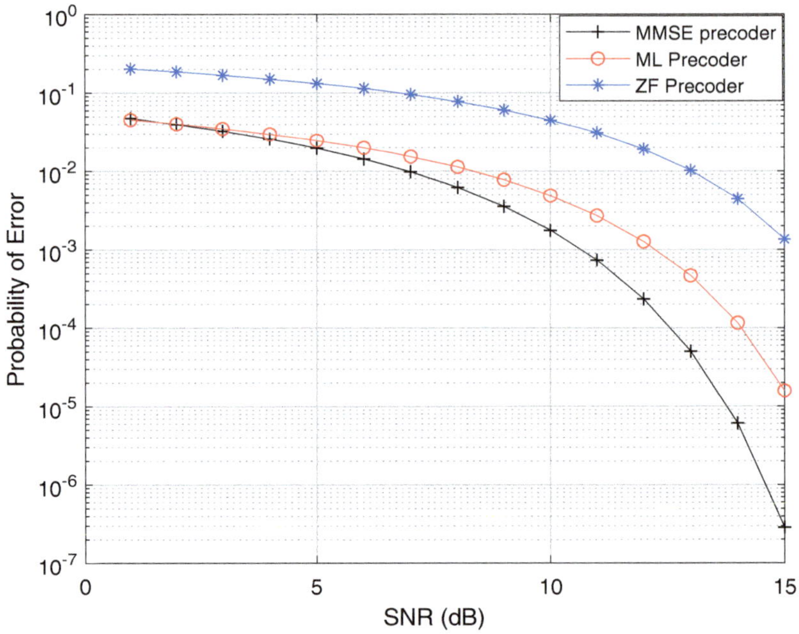

Figure 4.3. BER performance of suggested detection algorithms.

Figure 4.3 shows the comparison of the investigated schemes from BER standpoint. The MMSE precoding scheme shows better improvement in terms of BER. For MMSE and ML detectors, we can see here that the performance curves of these two systems are close to each other when SNR is low, but the gap gets larger when SNR gets higher, while the performance of ZF shows a gap with ML and MMSE for all values of SNR. For high SNR the gap is about 2 dB between MMSE and ML, while the gap between ZF and the two other schemes is about 8 dB. When SNR is less that means noise is large and the different detection schemes have almost the same behavior. When the SNR is large, the post detection of SNR may have been affected by channel matrix **H**.

Figure 4.4 shows a comparison of the suggested detection algorithms in terms of MSE. The MSE is a measure of the quality of an estimator, it is always non-negative, as shown in figure 4.4. However, the MSE values for all detection schemes are very close and very low which mean those schemes can be valid to use with the V2V–VLC system as values closer to zero are better.

In general, MMSE shows the best performance over ML and ZF because the MMSE detector is dependent on the noise variance and the condition number of the channel matrix.

4.5 Chapter summary

This chapter investigates the precoding and equalizing scheme for the V2V–VLC system considering flickering/dimming control and the nature of the outdoor VLC

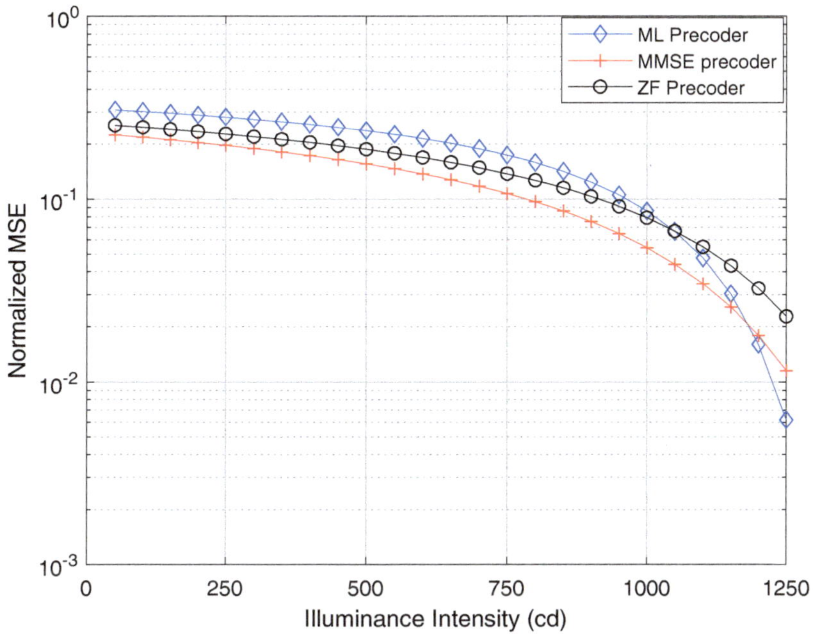

Figure 4.4. MSE performance of suggested detection algorithms.

channel as a frequency selective channel to enhance the system's performance. We present the precoder matrix, as well as the equalizer matrix. We compare the system performance of MMSE, ZF, and ML precoders, where the MMSE scheme shows a noticeable improvement.

References

[1] Tomaš B, Tsai H M and Boban M 2014 Simulating vehicular visible light communication: physical radio and MAC modeling *IEEE Vehicular Networking Conf. (VNC)* pp 222–5

[2] You J, Liu E, Wang P, Xu Z, Wang R, Gao Q and Hua Y 2017 Linear transceiver designs for MIMO indoor visible light communications under lighting constraints *IEEE Trans. Commun.* **65**

[3] Zeng L, O'Brien D, Minh H, Faulkner G, Lee K, Jung D, Oh Y and Won E T 2009 High data rate multiple input multiple output (MIMO) optical wireless communications using white LED lighting *Sel. Areas Commun.* **27** 1654–62

[4] Ko Y-C, Park K-H and Alouini M 2013 On the power and offset allocation for rate adaptation of spatial multiplexing in optical wireless MIMO channels *IEEE Trans. Commun.* **61** 1535–43

[5] Baxley Q H, Ying R J and Yao K S 2015 Joint optimization of precoder and equalizer in MIMO VLC systems *IEEE J. Sel. Areas Commun.* **33**

[6] Ma H, Lampe L and Hranilovic S 2013 Robust MMSE linear precoding for visible light communication broadcasting systems *IEEE Globecom Workshops (GC Wkshps)* pp 1081–6

[7] Ko Y-C, Park K-H and Alouini M S 2013 On the power and offset allocation for rate adaptation of spatial multiplexing in optical wireless MIMO channels *IEEE Trans. Commun.* **61** 1535–43
[8] Elgala H, Butala P M and Little T D C 2013 SVD-VLC: a novel capacity maximizing VLC MIMO system architecture under illumination constraints *Proc. of IEEE Globecom Workshop OWC*
[9] De W, Changyuan Y, Jian C, Chin F, Shin P, Wang C W and Zhong Z 2012 Performance of dimming control scheme in visible light communication system *Opt. Express* **20** 18861–8
[10] Walewski T, Sphicopoulos J, Ntogari T and Kamalakis G 2011 Combining illumination dimming based on pulse-width modulation with visible-light communications based on discrete multitone *IEEE/OSA J. Opt. Commun. Net.* **3** 56–65
[11] Arnon S 2015 *Visible Light Communication* (Cambridge: Cambridge University Press)
[12] Farahneh H, Hussain F, Hussain R and Fernando X 2018 A novel optimal precoder and equalizer in 2 × 2 mimo VLC systems for vehicular application *IEEE Globecom Workshops (GC Wkshps) (Dec 2018)* pp 1–6
[13] Luo Z-Q, Ding Y, Davidson T N and Wong K M 2003 Minimum BER block precoders for zero-forcing equalization *IEEE Trans. Signal Process.* **51** 2410–23
[14] Ozaki T, Iwai H and Sasaoka H 2014 Improvement of transmission performance by variable precoding in MIMO transmission *IEICE Commun. Express* **3** 168–74
[15] Wang R, Gao Q, You J, Liu E, Wang P, Xu Z and Hua Y 2017 Linear transceiver designs for MIMO indoor visible light communications under lighting constraints *IEEE Trans. Commun.* **65** 2494–508
[16] Davidson T N, Chan S S and Wong K M 2004 Asymptotically minimum BER linear block precoders for MMSE equalisation *IEEE Proc. on Communications* **vol 151** pp 297–304
[17] Fang D 2013 Optimal precoder design and block-equal QRS decomposition for ML based successive cancellation detection *Master Thesis McMaster University*
[18] Light emiting diode www.powerbulbs.com
[19] Photo diode parameters www.hamamatsu.com
[20] Farahneh H, Kamruzzaman S and Fernando X 2018 Differential receiver as a denoising scheme to improve the performance of V2V–VLC systems *ICC 2018-USA*

IOP Publishing

Visible Light Communications
Vehicular applications
Xavier Fernando and Hasan Farahneh

Chapter 5

Shadowing effects on visible light communication

Light is the most important idea behind visual representation of anything that a human being can visually perceive. The idea of perception of light lies in the fact that what you can see is not based on the objects that you are viewing but on the rays of light cast from a light source and reflected from those objects. It is important to note that your eyes do not directly see objects as there is no physical correlation between your eyes and those objects. All of this is theoretical, of course.

Light rays commonly originate from an energy source such as the Sun or a lamp in your room. It is important to note that theoretically a ray of light travels in a straight line and by the time you visually perceive an object, it is the rays of light reflected or scattered off that object that your eyes absorb.

Shadow is a dark (real image) area where light from a light source is blocked by an opaque object. It occupies all of the three-dimensional volume behind an object with light in front of it. The cross section of a shadow is a two-dimensional silhouette, or a reverse projection of the object blocking the light. An object is almost never simply in light and shade. Rather, it is usually in an environment in which light is bouncing around in several directions. For this reason, it is important for beginners to understand the nature of shadows and light.

5.1 Shadowing in VLC

Due to the nature of the visible light spectrum, shadowing is a significant issue, especially over short distances and small FOVs. This can seriously impair communications in both quasi-stationary indoor and fast varying outdoor environments.

However, by employing transmitters with a wide FOV as well as effectively utilizing the diffraction phenomenon of the light, the issues caused by shadowing can be alleviated. Note, it is much better to implement diffuse transmitters which radiate the light over a wide solid angle, which does not have as severe pointing and

doi:10.1088/978-0-7503-2284-3ch5

shadowing problems when compared to point-to-point links. The diffused transmitter does not need to be aimed directly at the receiver since the radiated optical waves reflect from multiple surfaces.

In this chapter, the effect of the shadowing on VLC system is discussed. Also, a method to take advantage of the optical diffraction phenomenon to overcome shadowing by employing a receiver with a wide FOV was proposed. Moreover, the shadowing effect for visible light was modeled by a bimodal distribution and the probability of error was derived for no, moderate and severe shadowing conditions. The bimodal distribution effectively represents separate propagation modes caused by different polarizations. This approach is useful in both indoor and outdoor environments [1].

5.2 Channel impulse response with shadowing

To mitigate the effect of shadowing, spatial diversity is commonly implemented where multiple transmitters and receivers are deployed around the shadowed area [2]. However, having such a setup is not always guaranteed and sustainable, henceforth it is necessary to study the effect of shadowing without having spatial diversity. This is necessary to effectively design a VLC system, along with obtaining the parameters that control its performance determined by the signal error probability under various shadowing conditions.

In this study, three cases of shadowing will be presented, no (absent), moderate and severe shadowing.

In a typical VLC system, the channel impulse response for a bimodal obstruction is

$$h_1(t) = \frac{\left(\frac{l+1}{2\pi}\right)A_r \cos^l(\phi)\cos(\theta)}{R_1^2}\delta\left(t - \frac{R_1}{c}\right) \tag{5.1}$$

$$h_2(t) = \frac{\left(\frac{l+1}{2\pi}\right)A_r \cos^l(\phi)\cos(\theta)}{R_2^2}\delta\left(t - \frac{R_2}{c}\right) \tag{5.2}$$

where l denotes the order of the source, R_1 and R_2 are the total distance from the lighting source to the receiver, taking into account the light diffraction around the obstructing object from the first edge and the second edge of the obstructing object, respectively, ϕ is the emission angle, θ is the incident angle, and c is the speed of the light.

The received signal energy E_r is,

$$E_r = \int_0^T |x(t) \otimes (h_1(t) + h_2(t))|^2 dt. \tag{5.3}$$

where $x(t)$ denotes the transmitted signal, T denotes the overall bit duration, and $\otimes$ means convolution sign.

5.3 Shadowing effect

In a standard VLC system without shadowing, it can be assumed that the transmitted signal encounters AWGN, hence the distribution at the receiver remains Gaussian. Since shadowing plays a significant role in VLC, the final distribution will result in a multi-mode distribution depending on the severity of the shadowing effect as well as the complexity of the obstructing object (i.e. non-uniform objects with complex cavities cause secondary interference effects thereby varying the channel gain) but a bimodal distribution will be considered here for simplicity.

As apparent in figure 5.1 (assuming the light only passes over two edges), a bimodal power distribution can be considered. With a bimodal approach, and assuming the PD receiver holds a large FOV, the optical power will be received from two distinct edges passing the object, each with different polarization.

Therefore, depending on the severity/complexity of the object surface, each polarizing path passing the object will appear as a local maximum.

Since we are considering the simple case of the light passing only from two sides of the object, this can be represented as a bimodal distribution with two peaks represent bit '1' and bit '0', respectively. The separation between the peaks of this distribution in a time-varying environment can be represented by two means and variances which both depend on time.

The distance between the means of the two normal functions which is roughly the distance between the two peaks is given by [3]

$$S(t) = \frac{\mu_1(t) - \mu_0(t)}{2(\sigma_0(t) + \sigma_1(t))} \tag{5.4}$$

where μ_0, σ_0, are the mean and the standard deviation of the first Gaussian distribution (bit '0'), μ_1, σ_1 are the mean and the standard deviation of the second

Figure 5.1. Different forms of shadow and light.

Gaussian distributions (bit '1'), and $\mu_1 > \mu_0$, $S(t)$ is a bimodal separation at a certain time.

As shown in figure 5.2, the bimodal distribution has the following PDF [4]

$$f(z, \mu_0, \sigma_0, \mu_1, \sigma_1) = \frac{1}{2\sqrt{2\pi(\sigma_0)^2}} \exp\left(\frac{-(z-\mu_0)^2}{2(\sigma_0)^2}\right) + \frac{1}{2\sqrt{2\pi(\sigma_1)^2}} \exp\left(\frac{-(z-\mu_1)^2}{2(\sigma_1)^2}\right) \quad (5.5)$$

Notice that when $S(t)$ decreases with a uniform ideal obstruction with minimal polarizing effects, the distribution will approach a standard Gaussian distribution and the received power will contain only a single maximum.

5.4 Error probability

In short-distance LOS links, a multipath dispersion is insignificant. Hence, LOS links channel are often only modeled with a linear attenuation and delay [5]. Optical LOS links are considered as non-frequency selective where the path loss depends on the square of the distance between the transmitter and the receiver [1].

In this study, the error probability derivation with shadowing effect under a bimodal distribution case assuming OOK modulation is employed, also, we assume that LEDs have a Lambertian radiation pattern [6]. This can, however, be extended to a multi-modal case.

To derive the probability of error, P_e of the received signal for the three cases of shadowing: no-shadowing, moderate and severe shadowing, considering figures 5.3–5.5 to illustrate their effect.

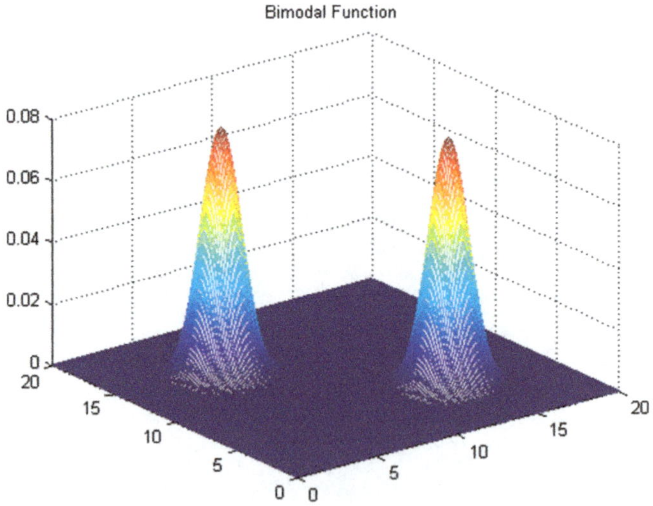

Figure 5.2. Bimodal function with equal variances.

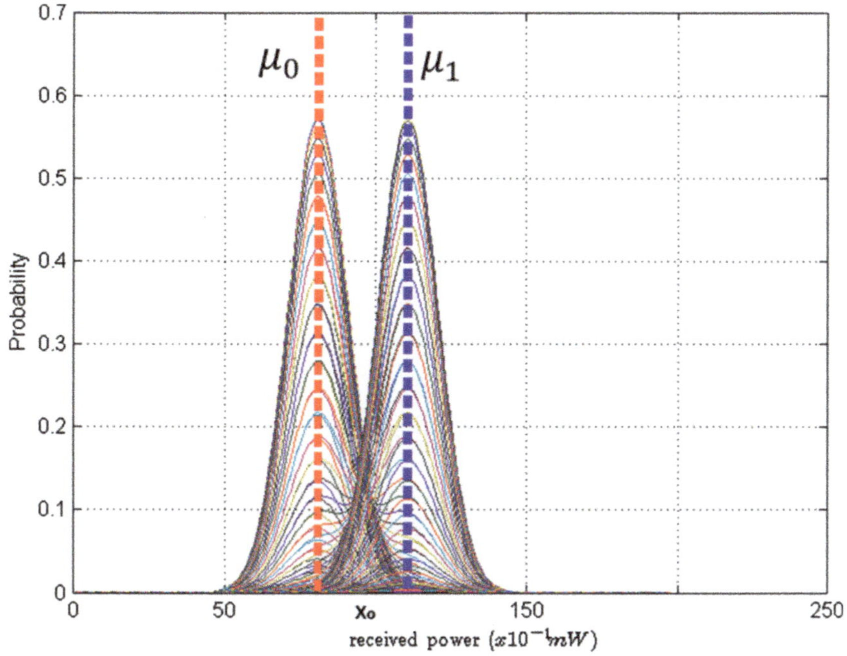

Figure 5.3. The PDF of no shadowing.

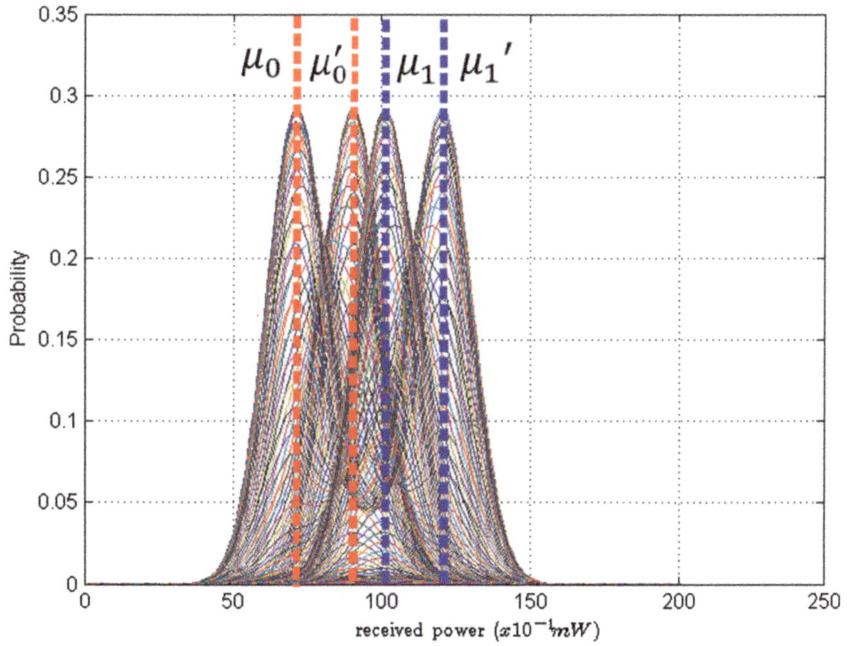

Figure 5.4. The PDF of moderate shadowing.

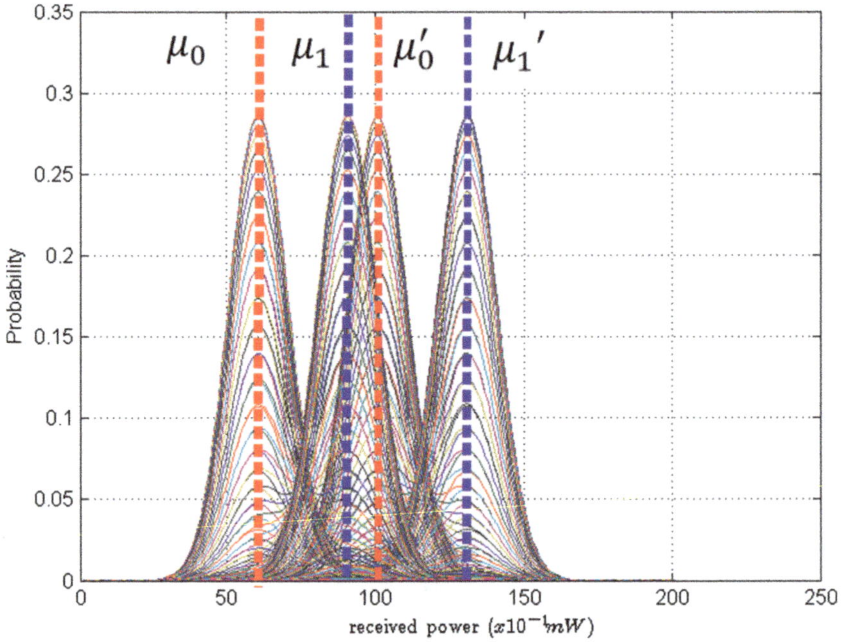

Figure 5.5. The PDF of severe shadowing.

In this study, a closed formula for the probability of error for each case of shadowing is given, hence, the derivation will be given for the first case, while, the other cases can be derived using a similar procedure.

It is important to note that in the case of no shadowing, the transmitted light will be received without being significantly deteriorated by the shadowing object, thus, the received signal will be modeled as a Gaussian distribution for both cases of transmitting 1 and 0 represented by s_1 and s_0, with average values μ_0, μ_1, respectively, as illustrated in figure 5.3.

In the case of having moderate or severe shadowing, the light will be more affected by the shadowing object, hence the signal will be received from both sides, which will result in a bimodal distribution. This is clearly depicted in figures 5.4 and 5.5, where two Gaussian distributions are created for each received bit, thereby resulting in a bimodal distribution.

The average expected received values can be represented by four vertical lines, with two possible average values for the case of receiving s_1, represented with average values μ_1, $\mu_1^{'}$ and two possible average values for the case of receiving s_0 represented by μ_0, $\mu_0^{'}$.

However, considering the fact that the shadowing is moderate, the two average values of receiving s_0 are close to each other, which also applies for the case of receiving s_1. Therefore, there are two decision regions for the probability of error calculation.

With respect to severe shadowing, the average values of the two modes are shifted significantly from each other. In this case, the shift is large enough to approximate the error by representing each mode as a normal distribution. This can be considered when the separation, $S(t) > 1$ between modes, then the distribution becomes two normal distributions, as shown in figure 5.5 [1].

Moreover, due to the severe shadowing case, the Gaussian distributions of s_0 and s_1 become close to each other which significantly affects the probability of error, which will be verified in the simulation section.

Furthermore, due to the obstruction effects, the width of each Gaussian distribution is halved when compared to the case of no shadowing due to the transmitted signal power being distributed between the two paths around the obstructing object.

In the following, the derivation of the probability of error P_e can be given as

$$P_e = P(0)P(e/0) + P(1)P(e/1) \tag{5.6}$$

where $P(0)$ and $P(1)$ are the probability of transmitting '0' and '1', respectively, $P(e/0)$ and $P(e/1)$ are the conditional probability for receiving '1' while '0' was transmitted and receiving '0' while '1' was transmitted, respectively, which is depicted by the shadowing in figure 5.3.

Due to symmetry, one can conclude that $P(e/0) = P(e/1)$.

As depicted in figure 5.3, the two Gaussian distributions have average values μ_0, and μ_1 representing the bits 0 and 1, respectively,

Define the decision region as [1]

$$x_0 = \frac{\mu_0 + \mu_1}{2} \tag{5.7}$$

where x_0 is the mid-point between the two means as shown in figure 5.3.

Then $P(e/0)$ is defined as

$$P(e/0) = \int_{xo}^{\infty} \frac{1}{\sqrt{2\pi\sigma^2}} \exp\left(\frac{-(x-\mu_0)^2}{2\sigma^2}\right) dx \tag{5.8}$$

Also, $P(e/1)$ is defined as

$$P(e/1) = \int_{-\infty}^{x_0} \frac{1}{\sqrt{2\pi\sigma^2}} \exp\left(\frac{-(x-\mu_1)^2}{2\sigma^2}\right) dx \tag{5.9}$$

Recall the Q function as

$$Q_x = \int_{x}^{\infty} \frac{1}{\sqrt{2\pi}} \exp\left(\frac{-y^2}{2}\right) dy \tag{5.10}$$

let $y = \frac{x-\mu_0}{\sigma}$, then, $dy = \frac{dx}{\sigma}$,

By Changing the integration limits, from (5.8), $P(e/0)$ can be given as

$$P(e/0) = \int_{\left(\frac{x_0-\mu_0}{\sigma}\right)}^{\infty} \frac{1}{\sqrt{2\pi}} \exp\left(\frac{-y^2}{2}\right) dy \tag{5.11}$$

then

$$P(e/0) = Q\left(\frac{x_0 - \mu_0}{\sigma}\right) = Q\left(\frac{\frac{\mu_0}{2} + \frac{\mu_1}{2} - \mu_0}{\sigma}\right) \tag{5.12}$$

Equation (5.12) can be simplified to

$$P(e/0) = Q\left(\frac{\mu_1 - \mu_0}{2\sigma}\right) \tag{5.13}$$

The $P(e/1)$ can be found by following same steps, where $P(e/1) = Q\left(\frac{\mu_1 - \mu_0}{2\sigma}\right)$.

Recalling (5.6), the probability of error for no shadowing will be given by

$$P_e = Q\left(\frac{\mu_1 - \mu_0}{2\sigma}\right) \tag{5.14}$$

To find the probability of error for moderate and severe shadowing when the intensity modulation is employed, the probability of error will be based on four average values due to the bimodal distribution, as shown in figures 5.4 and 5.5. Then by applying the same procedure, as shown above, the probability of error for moderate and severe shadowing can be given by

$$P_e = 2\left(Q\left(\frac{\mu_1 - \mu_0}{4\sigma}\right) + Q\left(\frac{\mu_1' - \mu_0}{4\sigma}\right)\right) \tag{5.15}$$

$$P_e = 4\left(Q\left(\frac{\mu_1' - \mu_0'}{4\sigma}\right) + Q\left(\frac{\mu_0 - \mu_1'}{4\sigma}\right) + Q\left(\frac{\mu_1 - \mu_0}{4\sigma}\right)\right) \tag{5.16}$$

Notice that in the case of moderate shadowing, we will still have two decision regions for detecting s_1 and s_0, while in the case of severe shadowing, the error probability is further increased, as shown before, due to the crossing of decision regions for detecting s_1 or s_0 which results in four decision regions.

5.5 Simulations

Matlab software was used to execute the simulation of the three cases of shadowing. A Gaussian channel was considered and the symbols s_1 or s_0 have equivalent variances but different means. Additionally, we are assuming intensity modulation is employed in the system where a stationary object with two main edges is obstructing the LOS signal, hence each modulation level will exhibit two modes. The error significantly increases when shadowing is in effect [1].

Figures 5.3–5.5 illustrate the density functions with the effects of shadowing. Here, blue and red lines represent modulation levels for s_1 and s_0, respectively.

Figure 5.3 illustrates a clear distinction between each modulated intensity level without exhibiting dual modes, hence there may be an obstruction with a uniform surface resulting in low shadowing or even no shadowing.

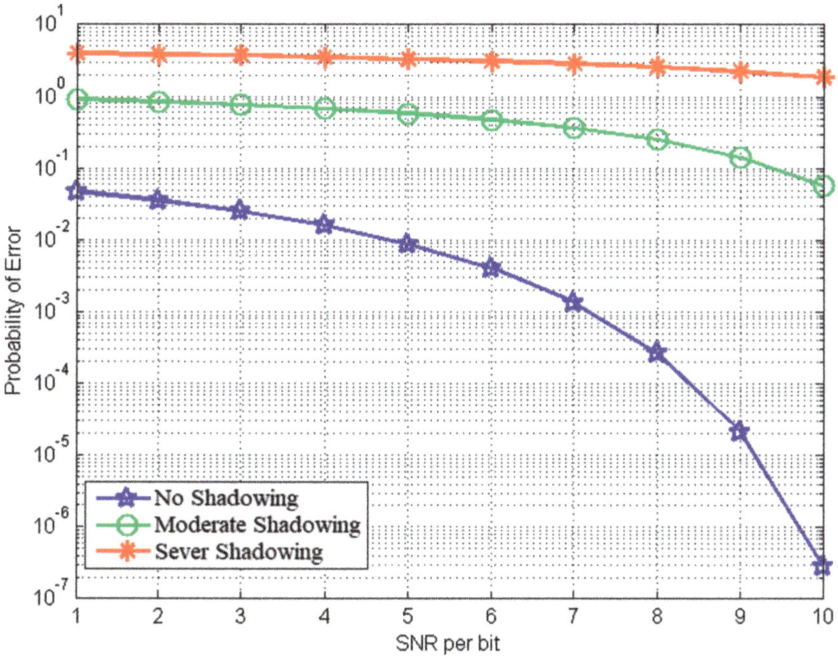

Figure 5.6. Error probability with different shadowing effects.

Figure 5.4 clearly shows the bimodal shadowing effect as there are two modes for each modulated intensity level.

In figure 5.5, the overlap between the respective modes resulting from severe shadowing becomes visible. It is shown that the respective intensity levels begin to overlap.

Shadowing becomes more severe as shown in figure 5.5. Hence, the decision region of bits 0 and 1 significantly overlaps, thereby greatly increasing the BER.

Also, we simulate the probability of error for different cases of shadowing by varying the received signal energies as well as its variances for the bimodal distribution. The simulation results clearly show that the error probability significantly increases due to the interleaving signal modes with each respective modulation.

Figure 5.6 depicts the result of our simulation which compares the error probabilities with different shadowing effects. Clearly, the severe shadowing has the worst error probability when compared to the no (absent) and moderate shadowing cases. It is also clear that for all shadowing cases, when the received signal power is increased, as a result of increasing the LED transmitted power, the probability of error decreases.

5.6 Chapter summary

In this chapter, the shadowing effect on the V2V–VLC system is studied. We propose a method to take advantage of the optical diffraction phenomenon to overcome shadowing by employing a receiver with a wide FOV. We model the shadowing effect for visible light by a bimodal distribution and have derived the probability of error for different types of shadowing.

References

[1] Farahneh H, Mekhiel C, Khalifeh A, Farjow W and Fernando X 2016 Shadowing effects on visible light communication channels *IEEE Canadian Conf. on Electrical and Computer Engineering (CCECE) (May 2016)* pp 1–5

[2] Zeng L, O'Brien D, Minh H, Faulkner G, Lee K, Jung D, Oh Y and Won E T 2009 High data rate multiple input multiple output (MIMO) optical wireless communications using white LED lighting *Sel. Areas Commun.* **27** 1654–62

[3] Brian E, Brian S, Zhang J and Mapes C 2003 Bimodality in tropical water vapour *Q. J. R. Meteorol. Soc.* **129** 2847–66

[4] Hastings N, Peacock B, Forbes C and Evans M 2005 *Statistical Distributions* 4th edn (New York: Wiley)

[5] Carruthers J B and Kahn J M 1997 Modeling of nondirected wireless infrared channels *IEEE Trans. Commun.* **45**

[6] Gfeller F R and Bapst U 1979 Wireless in-house data communication via diffuse infrared radiation *Proc. IEEE* **67** 1474–86

Chapter 6

Sunlight effect on V2V–VLC system and denoising schemes

Its well-known fact that 47% of the total solar irradiance falls within the visible light frequency band of the spectrum [1], as shown in figure 6.1. The main meteorological conditions in network locations determine the level of sunlight. Therefore, the major challenge of the outdoor VLC system is the strong influence of the ambient-light noise because of sunlight. sunlight represents unmodulated sources, that can be received at an average power much larger than the desired signal, even when optical filtering is employed. However, the noise due to the solar irradiance and other surrounding light sources, is a major concern that degrades the performance of the VLC system in outdoor application in terms of SNR and BER.

In this chapter, the effects of solar irradiance and other external sources are investigated for the V2V–VLC system with regard to SNR, BER, and data rate. Then, we presnt two schemes to combat the effect of sunlight on the V2V–VLC system. Firstly, we present the differential receiver as an efficient denoising scheme, then, we propose a kNN-algorithm- machine learning-based adaptive filter scheme. In the second scheme, our smart system can adapt itself according to varying noise conditions and help to achieve acceptable BER in support of reliable communications.

6.1 Solar irradiance

Solar irradiance is defined as the amount of solar power in any place. The earth's rate of rotation, the difference in local longitude, and the standard meridian for the local time zone are considered the main keys in irradiance calculations. Measuring irradiance correctly needs high accurate apparatus. In general, solar irradiance can be calculated from clear sky solar radiation on a horizontal surface in W m^{-2} [2].

The most important parameter to measure direct solar radiation (irradiance) is q factor and is given by [2]

doi:10.1088/978-0-7503-2284-3ch6

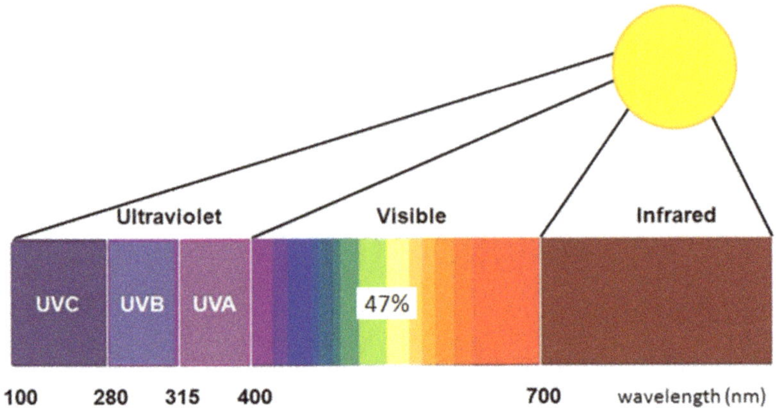

Figure 6.1. Spectrum of the sunlight.

$$q_{\text{sun}} = \left(1350.3 \left[1 + 0.099 \cos\left(\frac{360\, n}{365}\right) \right] [\sin \varphi \sin \zeta + \cos \varphi \cos \zeta \cos \omega] \right) \qquad (6.1)$$

where ζ is solar declination, ω is the angular displacement, φ is the location longitude, and n is the day of the year.

However, calculating solar irradiance is beyond the scope of this book, instead, the effect of solar irradiance on the V2V–VLC system will be addressed.

6.2 SNR of V2V–VLC system

In this study, we consider the system given in section 2.3. Also, we chose the city of Toronto as a reference to calculate solar irradiance during a full year (12 months) period to cover the worst and best cases of irradiance. We assume the sunlight incident angle is within the FOV of the receiver all the time. Furthermore, the OOK modulation scheme is used. The values of solar irradiance are given in table 6.1, while table 6.2 gives classes of natural light in different metrological conditions and at different times.

Figure 6.2 shows the scenario of solar irradiance effect on V2V–VLC, while figure 6.3 shows the response of the APD of predicted spectral irradiance of the sunlight for different visible light wavelengths in the city of Toronto. Here, APD response is given without any optical filter scheme. To show the strong effect of the solar irradiance on the V2V–VLC system, we calculate the SNR for the V2V–VLC system without any denoising scheme. figure 6.4 shows the strong effect of sunlight without any filtering scheme on SNR of the system during different months of the year. The summer months show the worst SNR, where SNR approaches $\simeq -3$ dBm, and in winter, SNR approaches $\simeq 9$ dBm.

6.3 Related work in denoising schemes for VLC

Most of the previous works for the outdoor-VLC system did not consider the effect of a practical ambient light noise which varies from time to time throughout the

Table 6.1. Solar irradiance at the city of Toronto [3].

Month	Average solar insolation Measured in kWh/m^2/day with solar panel direction southwest 45° from south
January	1.5
February	2.35
March	3.29
April	4.35
May	5.12
June	5.88
July	5.87
August	5.02
September	3.92
October	2.64
November	1.55
December	1.24

Table 6.2. Classes of natural light [4].

Natural Light Class	Intensity (lx)
sunlight not considered	0
Clear night, full moon	0.3
Winter day, overcast sky	900–2000
Summer day, overcast sky	4000–20 000
Winter day, clear sky	Up to 9000
Summer day, clear sky	Up to 100 000

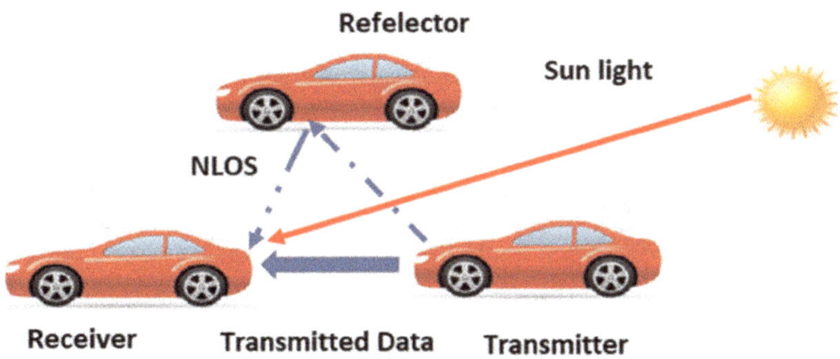

Figure 6.2. The effect of solar irradiance on V2V–VLC.

year, and from location to location. A few works have addressed the effect of sunlight in the outdoor-VLC system. In [1], the authors considered the impact of sunlight as a function of location, time and for different surfaces over the four seasons of the year and analyzed the solar irradiance. Moreover, they investigated

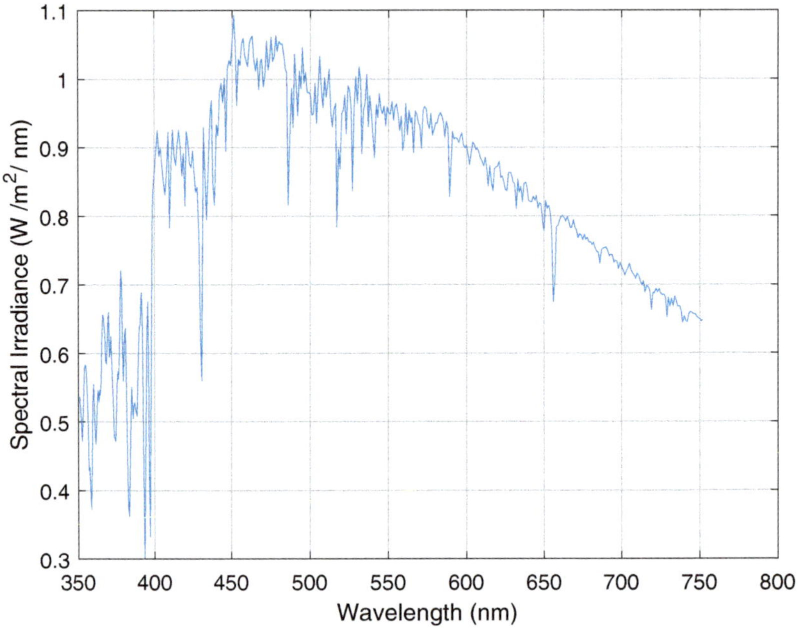

Figure 6.3. Average predicted solar irradiance in the city of Toronto in the month of June [3].

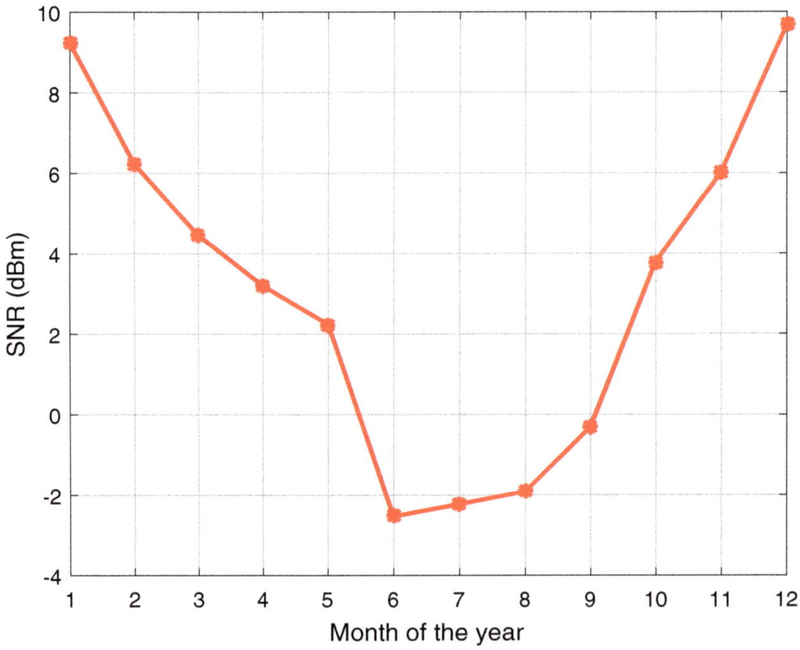

Figure 6.4. SNR without any filtration scheme i the city of Toronto—LED illuminance intensity is 1750 (cd) [3].

the effect of solar irradiance in the outdoor-VLC system in terms of data rate and BER degradation. The authors in [5] presented a daylight noise model based on a modified Blackbody radiation model to capture the effect of ambient-light noise. Also, they presented a new receiver using the selective combining technique to reduce the effect of background noise. The authors in [6] proposed a measurement-based time variant non-clear sky channel model for the I2V–VLC system which considers the dynamic characteristics of background radiation to enable more realistic and accurate prediction of the VLC system performance on the outdoor application. Moreover, they introduced a new receiver design with dual-reception and effective ambient-light rejection capabilities, which employs the selection diversity technique in order to mitigate the impact of ambient-light noise due to daylight. In [7], the authors presented an analytical analysis of solar irradiance. They investigated the effect of solar irradiance on the VLC system by using an optical filter in terms of data rate and BER degradation.

MLE has also been applied to the VLC field. The authors in [8] showed the applicability and feasibility of different MLE techniques based on iris recognition through smart phone captured images. The authors trained different classifiers and used histogram equalization processes to maximize accuracy. In [9], the authors proposed a novel object tracking framework based on an online learning scheme which can work robustly in challenging scenarios. They also proposed, a learning-based particle filter with color and edge-based features. Moreover, multiple classifiers fusion localization techniques using the received signal strengths of visible light are proposed in [10], in which LEDs are used to transmit various intensity modulated sinusoidal signals to be received by PDs placed at various grid points.

6.4 Noise calculations

At the receiver, the received signal consists of the transmitted signal and the noise signal. The photocurrent I_o due to the received optical power consists of two components, the LOS component and the NLOS component, i.e, ($I_o = I_{o_{LOS}} + I_{o_{NLOS}}$) and is given as

$$I_o = GA_r\left(\int_{\lambda_1}^{\lambda_2} P_r^{\mathrm{LOS}}(\lambda)\gamma(\lambda)T_o(\lambda)d\lambda + \int_{\lambda_1}^{\lambda_2} P_r^{\mathrm{NLOS}}(\lambda)\gamma(\lambda)T_o(\lambda)d\lambda\right) \tag{6.2}$$

where λ_1, $\lambda_2 \in [350\ 750](nm)$ are the visible light wavelengths, $P_r^{\mathrm{LOS}}(\lambda)$, $P_r^{\mathrm{NLOS}}(\lambda)$ are the received power and are given in (2.42), and (2.43), respectively, for certain wavelength λ, G is APD gain, $\gamma(\lambda)$ is the responsivity of the APD given in A/W, A_r is the effective area of the receiver, and $T_o(\lambda)$ is the transmittance of the bandpass optical filter.

Similarly, the photocurrent I_s produced at APD due to the solar irradiance is given as

$$I_s = GA_r\cos(\theta_{\mathrm{sun}})\int_{\lambda_1}^{\lambda_2} P_s(\lambda)\gamma(\lambda)T_o(\lambda)d\lambda \tag{6.3}$$

where $P_s(\lambda)$ is the solar irradiance given in W/m^2/nm, θ_{sun} is the incident angle of the sunlight on the receiver surface.

The random arrival of the incident photons from both LED light, and sunlight result in shot noise. This type of noise can be modeled by the Poisson process. Moreover, when the number of incident photons is large, shot noise is approximated by a Gaussian process. (using central limiting theorem).

According to [11] the shot noise variance is given as

$$\sigma^2_{\mathrm{Shot}} = 2q_e G^2 F(I_o + I_s)B \tag{6.4}$$

where q_e is the electron charge, F is the excess noise; $F = k_n G + (2 - \frac{1}{G})(1 - k_n)$ where k_n is the holes/electrons ionization rate [12], and B is the APD bandwidth.

6.4.1 Other noise sources

In addition to the shot noise, there are many different types of noises degrading the performance of the VLC system such as dark noise. Dark noise is caused due to avalanche effect of the APD and is given as [11]

$$\sigma^2_{\mathrm{Dark}} = 2q_e G^2 F I_{\mathrm{dg}} B + 2q_e I_{\mathrm{ds}} \tag{6.5}$$

where I_{ds} is the surface dark current and I_{dg} is the bulk dark current that experiences the avalanche effect of the APD.

Also, due to the random thermal motion of charge carriers, thermal noise is generated, the variance of thermal noise is given as [11]

$$\sigma^2_{\mathrm{Thermal}} = 4\left(\frac{K_{\mathrm{B}} T_{\mathrm{k}}}{R_{\mathrm{L}}}\right) F_{\mathrm{n}} \mathrm{B} \tag{6.6}$$

where K_{B} is Boltzmann constant, T_{k} is the temperature in Kelvin, R_{L} is the load resistance given as 50 Ω, and F_{n} is the photodiode noise figure.

Due to variances summation principle which is given in section 1.3, the total noise variance is given by

$$\sigma^2_n = \sigma^2_{\mathrm{Shot}} + \sigma^2_{\mathrm{Dark}} + \sigma^2_{\mathrm{Thermal}} \tag{6.7}$$

In this study, we consider Hamamatsu *S*8664 model for APD [13]. The parameters and their values are given in table 1.3.

6.4.2 Performance metrics

The SNR for the system is

$$\mathrm{SNR} = \frac{I_o}{\sigma^2_n} \tag{6.8}$$

The maximum achievable data rate defined as

$$R_{\max} = B \log_2(1 + \mathrm{SNR}) \tag{6.9}$$

For OOK modulation scheme, BER for certain SNR is given by (2.48).

6.5 Differential receiver as a denoising scheme to improve the performance of V2V–VLC systems

Achieving high SNR, high data rate, and minimizing multipath dispersion are the greatest challenges during the receiver design for any system. In order to design an efficient optical filtering system for the outdoor V2V–VLC system in the presence of sunlight and other artificial lights, eye safety should additionally be taken into account. Many previous works considered a bandpass blue optical filter for the denoising process [7, 14]. Such schemes include APD with collimate lens on the top of the APD to focus incident light onto the surface of the APD. The optical filter is laid on this lens for blocking unwanted lights.

In this study, a differential receiver with optical filter scheme is presented as a solution to combat the effect of the sunlight noise. We investigate the system performance for three scenarios: without optical filter, with optical filter, and with the proposed differential receiver.

6.5.1 Differential filtering scheme for 2 × 2 MIMO-V2V–VLC

The differential filtering scheme is a well-known idea in the communications field [15, 16]. In this study, we apply this scheme with some modifications to the system to mitigate the effect of sunlight noise. The differential filtering circuit is shown in figure 6.5. Each transmitter source transmits with different wavelength, i.e., transmitter 1 uses λ_1 and transmitter 2 uses λ_2, where the difference between λ_1 and λ_2 is several tens of nanometers, thus, we can assume the channel coefficients for two transmitters are equal. We consider red shift phenomena to adjust the wavelengths λ_1 and λ_2 [17].

For 2 × 2 MIMO, the received signal at any receiver $\mathbf{y}_{\mathbf{ri}}$ is givne by

$$\mathbf{y}_{\mathbf{ri}} = [h_{i1}\ h_{i2}][\mathbf{x}_1\ \mathbf{x}_2]^{\mathrm{T}} + \mathbf{n}_{\mathbf{i}} \tag{6.10}$$

where ($i = 1, 2$), $\mathbf{x}_1$, $\mathbf{x}_2$ indicate the transmitted vector of first and second transmitter, respectively, with wavelength λ_1, and λ_2. $\mathbf{n}_{\mathbf{i}}$ indicates the noise vectors including shot noise, which is rising up from the sunlight in addition to thermal noise and dark noise. We will assume only the shot noise, as it is the dominant noise.

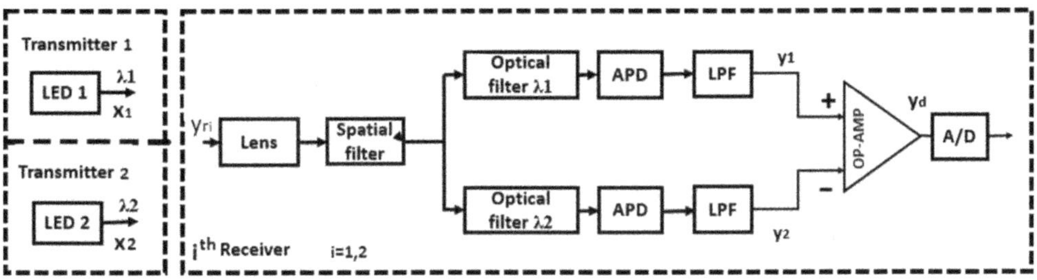

Figure 6.5. Differential denoising scheme for VLC.

The received signal is collected by a lens, as shown in figure 6.5. Thereafter, the signal passes to a spatial filter to alter the structure of light. Then, the signal passes through two parallel bandpass optical filters with equal bandwidth but different centre frequencies ($f_1 = \frac{c}{\lambda_1}, f_2 = \frac{c}{\lambda_2}$), where c is the speed of light. Afterwards, the outputs of the optical filters pass to APDs and then the signal passes through a low pass filter (LPF). At the end, two parallel signals $\mathbf{y}_1$, and $\mathbf{y}_2$ are subtracted by an operational amplifier (OP-AMP) to form one signal ($\mathbf{y}_d = \mathbf{y}_1 - \mathbf{y}_2$) to pass to an analog-to-digital (A/D) converter, and finally to the signal detection process [18].

Since the optical filter will pass either $\mathbf{x}_1$, or $\mathbf{x}_2$ depending on the signal wavelength, $\mathbf{y}_1$, and $\mathbf{y}_2$ are given by

$$\mathbf{y}_1 = \mathbf{H}_\mathrm{T}\mathbf{x}_1 + \mathbf{n}_1 \tag{6.11}$$

$$\mathbf{y}_2 = \mathbf{H}_\mathrm{T}\mathbf{x}_2 + \mathbf{n}_2 \tag{6.12}$$

where $\mathbf{H}_\mathrm{T}$ is the transfer function of the channel of one branch of the differential receiver. In this study, $\mathbf{H}_\mathrm{T}$ can be considered the same for two parallel branches.

The output of the OP-AMP, $\mathbf{y}_\mathrm{d}$ is given by

$$\begin{aligned} \mathbf{y}_\mathrm{d} &= \mathbf{y}_1 - \mathbf{y}_2 \\ &= \mathbf{H}_\mathrm{T}\mathbf{x}_\mathrm{d} + \mathbf{n}_\mathrm{d} \end{aligned} \tag{6.13}$$

where ($\mathbf{x}_\mathrm{d} = \mathbf{x}_1 - \mathbf{x}_2$), and ($\mathbf{n}_\mathrm{d} = \mathbf{n}_1 - \mathbf{n}_2$).

6.5.2 Signal filtration by differential receiver

Assuming the sunlight noise is a stochastic stationary random process, the background noise component after the LPF will be cancelled due to the differential detection system.

The sunlight noise spectrum, whose envelope denotes by S, is spread over a very large spectral band B_s. Because we assume each headlight transmits the signal with different λ, we need two band pass filters (BPFs) (optical filters) at the suggested differential detector with central frequencies f_1 and f_2. Their impulse response functions are $h_1(t)$ and $h_2(t)$, respectively, and the corresponding transfer functions are $H_1(f)$ and $H_2(f)$, respectively. Also, let the two filters have same bandwidth given by B_f. Since $B_s \gg B_f$, the output of the filters S_1 and S_2 can be considered flat.

Assuming the optical filter is a linear time-invariant system, S_1 and S_2 are given by [18]

$$S_i(t) = \int_{-\infty}^{\infty} h_i(\tau)S(t - \tau)d\tau \quad i = 1, 2. \tag{6.14}$$

The filters output S_1 and S_2 are passed to the APD. At the output of the ith APD, the light intensity is given by

$$I_{Si}(t) = \int_{T_s} |S_i(t)|^2 dt \tag{6.15}$$

where T_s is the symbol period of the assigned modulation scheme.

Hence, $|S_1(t)|^2$ is given by

$$|S_1(t)|^2 = \int_{-\infty}^{\infty}\int_{-\infty}^{\infty} h_1(\tau_1)S(t-\tau_1)h_1(\tau_2)S(t-\tau_2)d\tau_2\, d\tau_1 \tag{6.16}$$

For the first APD, we substitute (6.16) into (6.15), and get

$$I_{S1}(t) = \int_{-\infty}^{\infty}\int_{-\infty}^{\infty} h_1(\tau_1)h_1(\tau_2)\int_{T_s} S(t-\tau_1)S(t-\tau_2)d\tau_2\, d\tau_1\, dt \tag{6.17}$$

Since S is considered to be an ergodic stationary random process, then the third integral in (6.17) is the autocorrelation function of $S(t)$. Thus, we can write it in terms of expected value $E[.]$ as

$$E(S(t-\tau_1)S(t-\tau_2)) = \int_{T_s} S(t-\tau_1)S(t-\tau_2)dt = R_s(\tau_2-\tau_1) \tag{6.18}$$

Assuming that H_1 is an ideal BPF, its impulse response is $h_1(t) = \text{sinc}(2\pi B_f t)\cos(2\pi f_1 t)$. The impulse response $h_1(t)$ is a sinc function with its first zero at $\frac{1}{\Delta f}$ $(\Delta f = f_2 - f_1)$, modulated by $\cos(2\pi f_1 t)$. On the other hand, assuming a rectangular-shaped power spectral density (PSD) for S (just for the sake of demonstration simplicity), its autocorrelation function will be a sinc function with its first zero at $\frac{1}{B_S}$. Given that $B_s \gg \Delta f$ in practice, this autocorrelation can relatively be considered as a Dirac delta function. In other words, S can be considered as a white noise in the calculation of I_{S1}. Figure 6.6 shows an illustration for the PSD of the background noise. Also, it shows bandpass filtering of bandwidth B_f around f_1 and f_2, corresponding to the bandpass background noises' field envelopes S_1 and S_2.

The autocorrelation function in (6.18) can be given by

$$R_s(\tau_2-\tau_1) = \sigma_s^2\delta(\tau_2-\tau_1) \tag{6.19}$$

where σ_s^2 is the variance of the noise and δ is Dirac delta function.

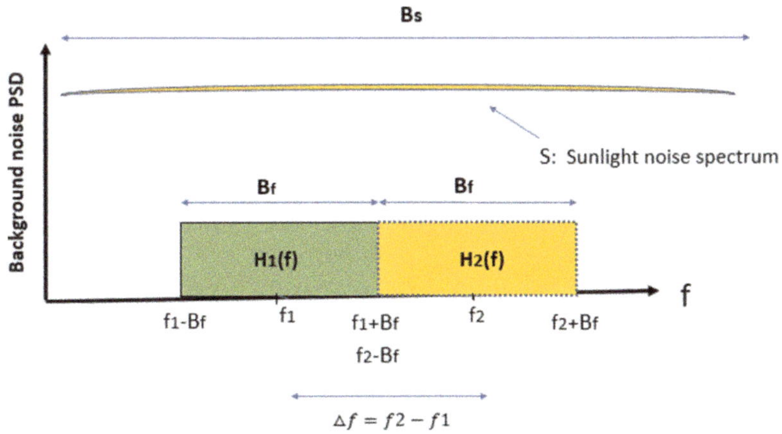

Figure 6.6. PSD of the background noise.

Furthermore, (6.15) can be written as

$$I_{S1}(t) = \sigma_s^2 \int_{-\infty}^{\infty} h_1^2(\tau)d\tau \tag{6.20}$$

Considering both filters are identical, at the operational amplifier (OP-AMP), recall figure 6.5, two light intensities I_{S1} and I_{S2} will be subtracted, thus

$$I_{S1} - I_{S2} = \sigma_s^2 \int_{-\infty}^{\infty} \left[h_1^2(\tau) - h_1^2(\tau)\right]d\tau \tag{6.21}$$

This is what we had denoted by n_d in (6.13). Using the Parseval equality, we obtain

$$n_d = \sigma_s^2 \int_{-\infty}^{\infty} [|H_1(f)|^2 - |H_2(f)|^2]df = 0 \tag{6.22}$$

Equation (6.22) indicates that the shot noise, which rises by the sunlight, in V2V–VLC can be neglected by using a differential detector.

6.5.3 Simulation results

We use Matlab to simulate the proposed system and consider the city of Toronto for calculating solar irradiance. Simulation parameters are given in tables 1.3, 2.1, 3.1, and 6.1 [18].

Figure 6.7 shows the response of the APD of the predicted spectral irradiance of the sunlight for different visible light wavelength. Here, APD response is given for three cases: without optical filter, with blue optical filter, and with differential receiver. Moreover, figure 6.7 shows that the violet and blue lights have less irradiance, while yellow and green lights have the highest irradiance. A differential receiver captures up to 60% of solar irradiance while a blue filter captures up to 75%, which means the proposed scheme shows good improvement in the system performance.

The proposed system performance is studied by investigating the average BER versus the sunlight irradiance for different months of the year for the three scenarios, as shown in figure 6.8. A significant improvement is achieved when the differential receiver and blue filter are used. The differential receiver can improve the BER up to $\simeq 5 \times 10^{-5}$, while a blue filter scenario can improve the BER up to $\simeq 1 \times 10^{-4}$.

Figure 6.9 shows the data rate versus the sunlight irradiance for the three considered scenarios. It is shown that the proposed scheme improves the data rate three times compared to without any filtering scheme, and by one and half times compared to a traditional filtering scheme. It is also shown by figures 6.8 and 6.9 that the V2V–VLC system can still work under solar irradiance, where the worst achievable BER during June (highest irradiance value) is $\simeq 5 \times 10^{-4}$.

6.6 V2V–VLC denoising scheme using machine learning

Machine learning (MLE) is a powerful sub-field of artificial intelligence and is being used for model training purposes in data mining, pattern recognition, and image

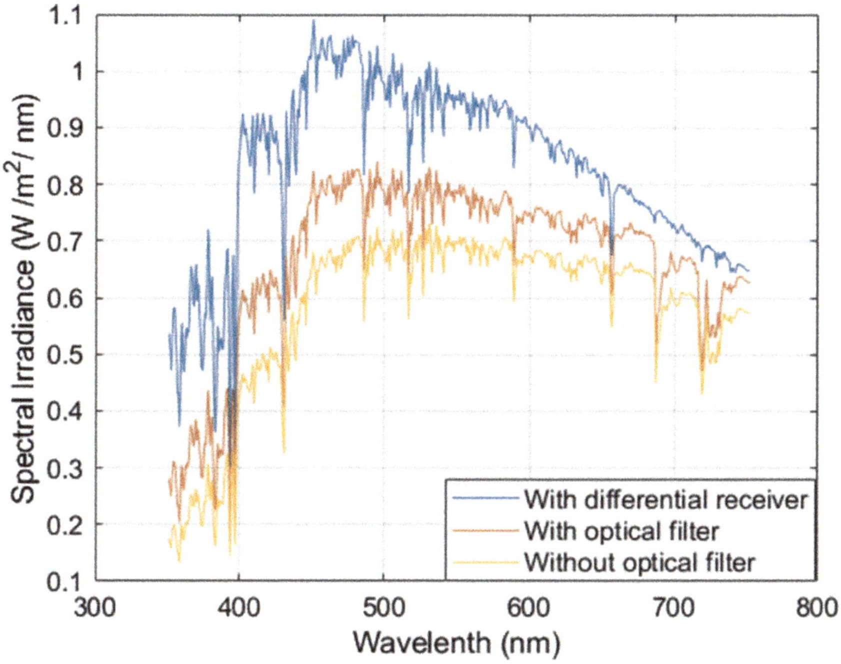

Figure 6.7. Average predicted solar irradiance at the city of Toronto in the month of June with denoising schemes.

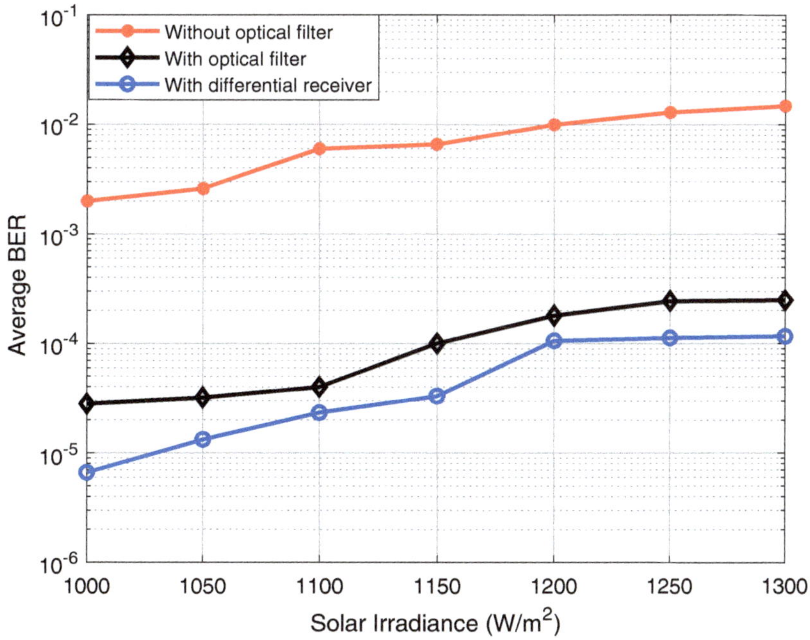

Figure 6.8. BER versus sunlight irradiance at the city of Toronto.

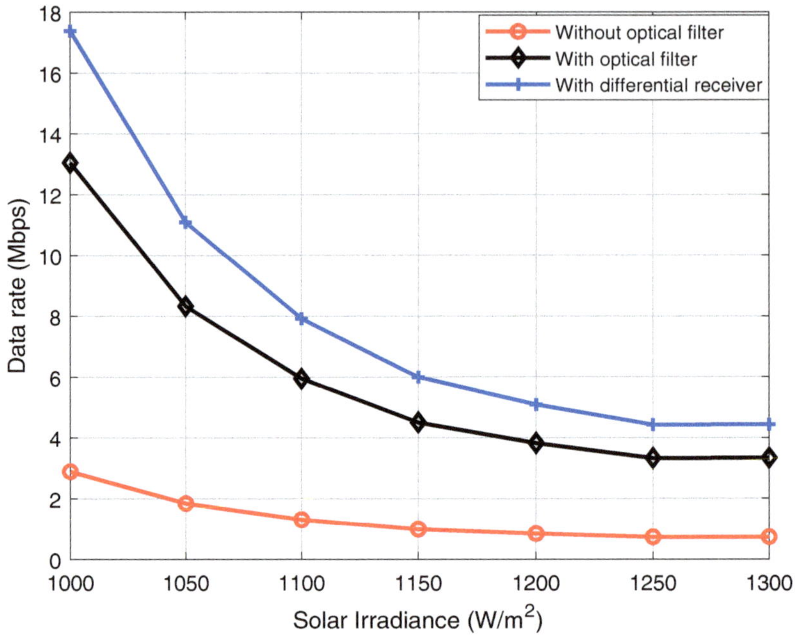

Figure 6.9. Achievable data rate versus sunlight irradiance at the city of Toronto.

processing. Centralized and distributed algorithms are required for data fusion and automated decision making during smart sensing/monitoring. MLE is used to develop technologies for machines/devices, for monitoring present behavior and predicting future behaviors. It has also been utilized in optical communication systems for indoor localization and failure prediction.

MLE techniques are becoming pervasive and they can play a key role in all stages of smart VLC networks–from data analysis to behavior prediction. In the second denoising scheme, we use kNN supervised learning technique, which is a simple and powerful algorithm intended to solve classification and clustering problems to combat the solar irradiance effect on the V2V–VLC system. It is a straightforward and effective method to find and indicates associations within a given data set.

6.6.1 System model

In this study, we consider the same model, which is given by section 2.3 and shown in figure 6.2, to apply MLE for performance enhancement of VLC-based V2V systems under a sunlight effect. Moreover, we have chosen the city of Toronto as a reference location for consideration of solar irradiance during the summer season. We consider the solar irradiance listed in table 6.1, and OOK modulation scheme for the VLC signal.

6.6.2 Machine learning and adaptive filtering

We apply an MLE algorithm for the training of an optical filter. This filter is installed within the taillights (the receiver) of the receiving vehicle. This filter adaptively modifies itself responding to physical conditions of the VLC link-attempting to achieve an acceptable BER.

Depending on the environmental conditions, the values for solar irradiance will change with weather and overtime day. The distance between two communicating vehicles will also vary due to the relative speeds between them. The FOV is expected to change with vehicle size and make, style and width of headlights and taillights. All of these factors affect the reception of the VLC signal and, in turn, affect the received BER.

We train our system using supervised learning. We use the nearest neighbour algorithm, which belongs to the 'Supervised Learning' MLE class.

6.6.3 kNN algorithm

The kNN algorithm is one of the simplest classification algorithms and it is one of the most used learning algorithms. Also, kNN is a non-parametric, easy learning algorithm. Its purpose is to use a database in which the data points are separated into several classes to predict the classification of a new sample point.

6.6.3.1 Advantages

- Simple to implement;
- Flexible to feature/distance choices;
- Naturally handles multi-class cases;
- Can do well in practice with enough representative data.

6.6.3.2 Disadvantages

- Large search problem to find nearest neighbours;
- Storage of data;
- Must know we have a meaningful distance function;
- The accuracy of kNN can be severely degraded with high-dimension data because there is little difference between the nearest and farthest neighbor.

In our study, we use kNN because we are dealing with two variables only (FOV, and solar irradiance) so we do not need too much storage data or high dimension data, thus, the calculation time will be reasonable. Also, we suggest a weighed voting factor to remedy skewed class distributions.

6.6.4 Problem classification

We train the system using the nearest neighbor supervised MLE algorithm. We use data for the city of Toronto for various values of solar irradiance which is shown in table 6.1. After developing the data set for various parameters affecting the VLC signal, the nearest neighbor algorithm was applied to find the closest match of the actual transmitted VLC signal. The suggested algorithm finds the closest neighbor to the transmitted signal and finds the effective BER. If this BER is over a certain threshold, the suggested system modifies itself to achieve the desired acceptable rate.

6.6.5 Adaptive filtering

We do not give details of the filter design in this book, but leave it for the future work. There are many factors that can effect BER (transmitted power, incident angle, received angle, distance, solar irradiance), but for now, we consider only two parameters, i.e, FOV, and solar irradiance. We train our system using kNN, and it will adjust itself according to the difference between achieved and desired BER. The system can perform two actions, depending on the received VLC signal and achieved BER.

- Blocking solar irradiance: we assume our filter is comprised of cascading films/blinds and can automatically make these films appear before APDs, for irradiance noise filtering from the VLC signal. The filter will adjust the number of films/blinds according to the intensity of solar irradiance. If the effect of solar irradiance is greater, the received BER is expected to be worse, therefore, the smart filter will automatically add more blinds to nullify the effect of ambient noise.
- Modifying the field of view angle: FOV can affect the VLC signal, as the incident angle of the received signal changes with different inclinations of taillight mounts. Changes in the incident angle will change the received signal strength and affect the BER. If achieved BER is not acceptable, our system will modify the inclination angle of taillights for proper FOV to achieve a better BER. The sequence of operations for our algorithm is given in figure 6.10.

6.6.6 k-Nearest neighbour algorithm and distance weighted kNN rule

kNN, is a well-known algorithm used in the pattern recognition literature. According to this algorithm, an unclassified pattern (sample, instance) is assigned to the class represented by a majority of its k-nearest neighbours. This rule is usually called the voting kNN rule. The number N of patterns and k are related such that, $K/N \mapsto 0$, the error rate of the kNN rule approaches the optimal Bayes error rate.

kNN regression is a nonparametric regression method, where the information derived from the observed data is applied to forecast the number of predicted variables in real time without defining a predetermined parametric relation between predictor and predicted variables. Also, in voting kNN, the k neighbours are implicitly assumed to have equal weight in the decision, regardless of their distances to the observed data x to be classified, or, otherwise, the weight should be define for each neighbour. The basis of this method is on calculating the similarity (neighborhood) of the real-time number of predictors $X_r = x_{1r}, x_{2r}, x_{3r}, \dots, x_{mr}$ with the amount of predictors for each historical observation $X_t = x_{1t}, x_{2t}, x_{3t}, \dots, x_{mt}$ via Euclidean distance function D_{Ec} or any other distance method. In this study, we use two methods, Euclidian distance and city block method (Manhattan distance) in search of a comparison for the performance.

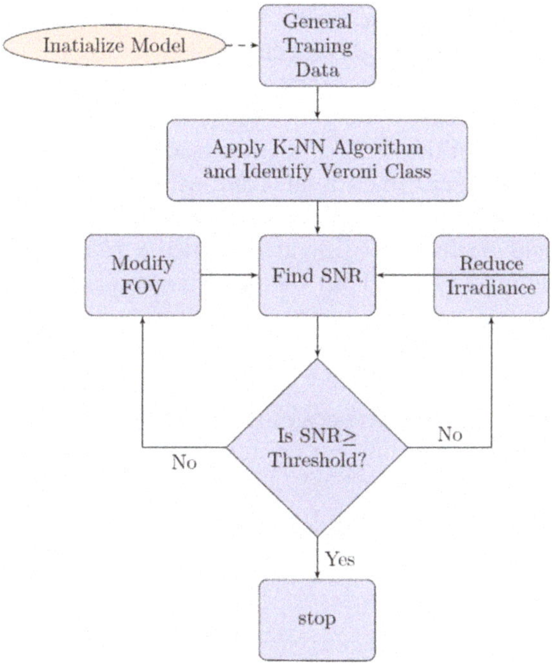

Figure 6.10. Adaptive filtering using the kNN algorithm.

Euclidian distance is given as follows

$$D_{Ec} = \sqrt{\sum_{i=1}^{m} w_i(x_{ir} - x_{it})^2} \quad t = 1, 2, 3, \ldots, n \tag{6.23}$$

where $w_i (i = 1, 2, \ldots, m)$ are the weights of the predictors, summation of which is equal to one.

The city block distance between two points is calculated as

$$E_{\text{bo}} = \sum_{i=1}^{K} |(x_{ir} - x_{it})| \tag{6.24}$$

In this study, the weight w_i is calculated as follows

$$w_i = \begin{cases} \dfrac{d(x_k, x) - d(x_i, x)}{d(x_k, x) - d(x_1, x)} & \text{if } d(x_k, x) \neq d(x_1, x) \\ 1 & \text{if } d(x_k, x) = d(x_1, x) \end{cases} \tag{6.25}$$

where $d(x_k, x)$ is the distance between the kth neighbor and the observation x.

More formally, given a positive integer K, an unseen observation x and a similarity metric distance, the kNN classifier performs the following two steps:

Step A : distance between data points For a certain value of K, the distance between observed data and its neighbors can be calculated using different methods such as Minkowsky, Euclidian, city block, etc. Afterwards, measured distances are sorted to determine the nearest neighbor based on the kth minimum distance. Thereafter, the categories of the nearest neighbors are gathered to do the voting. The algorithm runs through the entire data set, computing the distance D between x and each training data element. K points in training data closest to x are saved in set A.

Step B : Categorial Assignment For an instance x (observed data), assume we have N_i instances belong to class Y_i in the neighborhood. Then we define

$$P(Y_i/x) = \frac{N_i + s}{K + Cs} \tag{6.26}$$

where K is the total number of instances in the neighborhood, C is the total number of classes, and s is the smoothing parameter. The smoothing is used to avoid 0 probabilities.

A detailed self-explanatory description of our proposed algorithm is given in algorithm 1.

6.6.7 Simulation results

In the simulation, we use the MLE toolbox in the Matlab suit of software. The simulation parameters are shown in tables 1.3, 2.1, 3.1, and 6.1.

Algorithm 1: K nearest neighbour

Classify (Xi, Ci, x); where $i = 1, 2, 3, \ldots, K$ be the data points, X: feature values, C: class labels of Xi for each i, x is the unknown sample
BEGIN
for $i = 1$ **to** $i = K$ **do**
  Compute distance $d(Xi, x)$, where d denotes the distance between the two points
end for
Compute set A, containing indices for K smallest points $d(Xi, x)$
return majority label for Ci where $i \in$ A.

6.6.7.1 Data sets and classes

In figure 6.11, the Voronoi diagram is shown to represent the training and the observed data and their distribution to various classes on the basis of calculated BER. We plot data sets as a Voronoi diagram to make it easy to recognize the boundaries of decisions areas. The classification of data sets is based on the classes of its nearest neighbours. Since each data set consists of two variables: solar irradiance and FOV, the x-axis represents the irradiance and the y-axis represents the FOV.

The functionality of the kNN algorithm can be explained by considering one element of the data set. We denote the observed data in Voronoi, with a small circle

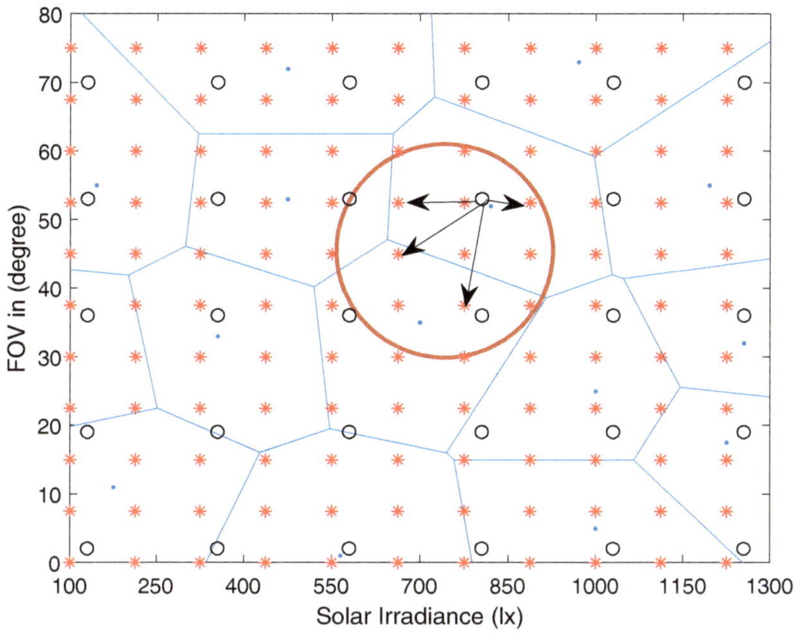

Figure 6.11. Training data and observed data in a Voronoi diagram.

'o', and the training data by small '*', for a certain observed data point. For a certain value of k, the distance between observed data and its neighbors can be calculated. Afterward, measured distances are sorted to determine the nearest neighbor based on the kth minimum distance.

6.6.7.2 Choice of k

K in kNN is the number of instances that we take into account for the determination of affinity with classes. So we need to investigate the performance of kNN near rule-of-thumb-value and make a decision about the optimal one using any algorithm for performance testing. In figure 6.12, the effect of the numbers of selected neighbors on the decision accuracy is presented. We use the Euclidian distance formula and City block distance formula to generate this figure, and use a Monte Carlo method to average it over 2000 iterations. It is shown that optimal value of K is 12 with Euclidian distance and K is 6 with city block distance.

6.6.7.3 BER discussion

Figure 6.13 represents the BER of the proposed system before applying the denoising scheme and after applying the scheme. Figure 6.13 shows improvement in terms of BER after applying the denoising scheme. We can notice an improvement in BER from 10^{-2} to 10^{-5}, thus, reflecting the stability and quality of the communication link.

Figure 6.14 shows the probability of error of each action (add optical filter or change the FOV) at each daytime hour. Because choosing the action is random, this

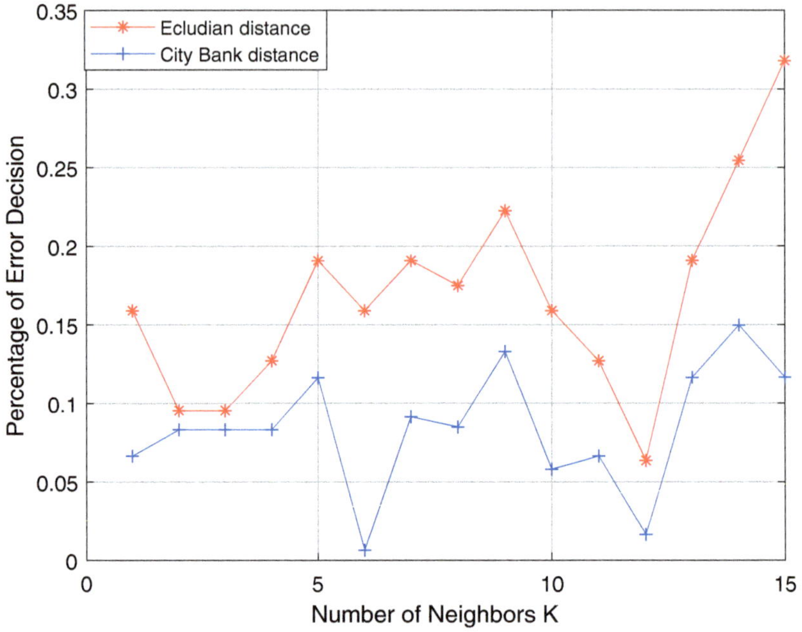

Figure 6.12. Variation of the number of neighbors with the percentage of decision error.

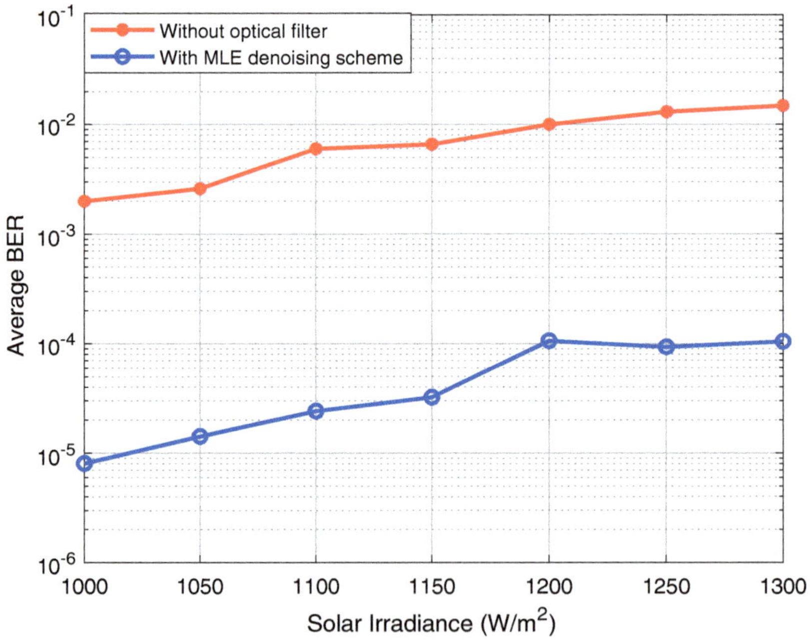

Figure 6.13. BER of the V2V–VLC system with and without the denoising scheme.

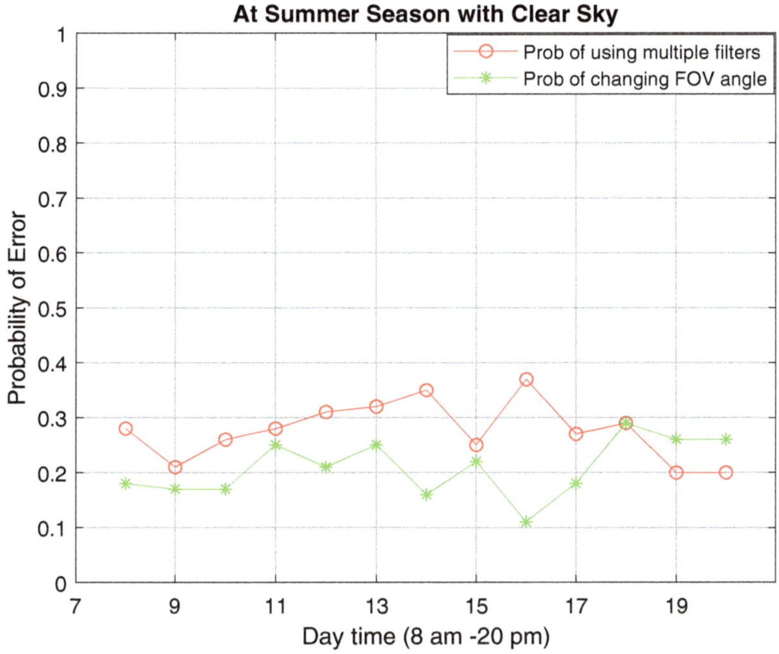

Figure 6.14. Hourly probability of noise reduction actions.

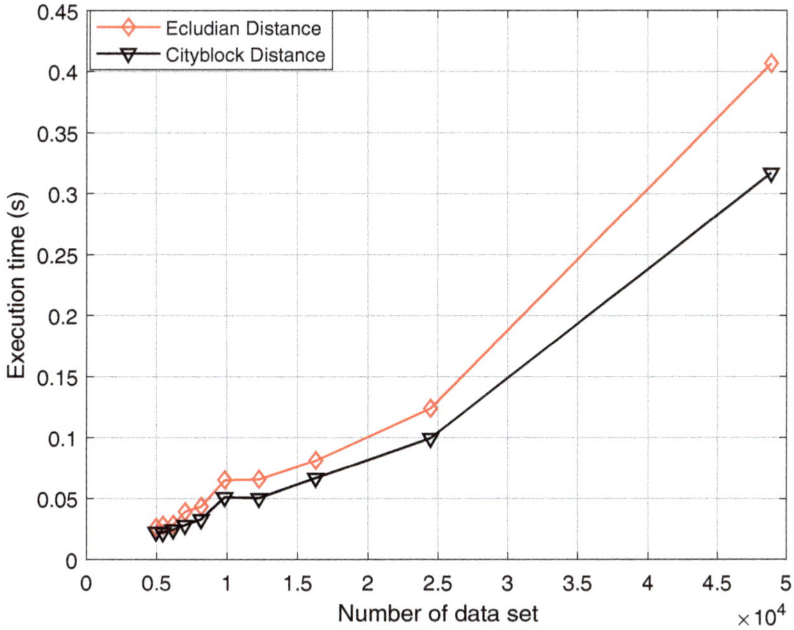

Figure 6.15. Execution time of the proposed scheme.

result may help to determine which action of the system has to be chosen at each hour of the day. Moreover, figure 6.14 shows the probability of adding an optical filter is higher than changing the FOV; the reason is that while times change, the position of the Sun also changes and the irradiance value will change.

6.6.8 Execution time

In figure 6.15, we discuss the execution time for the proposed scheme. We consider two methods to calculate the distance between the instance and its neighbors, Euclidian distance, and city block distance. Also, we use Intel(R) Core (TM) i5-5200U CPU@ 2.20 GHz laptop to calculate the execution time for the data sets. Figure 6.15 shows both methods almost have the same time of execution for a low numbers of data sets, 0.05 s for a 5000 data set, while the city block method is better for a high number of data sets, 0.32 s for a 50 000 data set.

6.7 Chapter summary

This chapter proposes two denoising schemes to combat the effect of solar irradiance, thus improving the performance of V2V–VLC systems under sunlight effect. The first proposed denoising scheme is by using a differential receiver, while the second proposed scheme is by using a kNN algorithm as a machine learning based adaptive filter. Simulation results of both schemes show good improvement in SNR, which makes the communication link more stable even in the months of summer.

References

[1] Beshr M, Andonovic I and Hussien M 2012 The impact of sunlight on the performance of visible light communication systems over the year *Proc. SPIE* **8540** 85400F

[2] Duffie J A and Beckman W A 2013 *Solar Engineering of Thermal Processes* (New York: Wiley)

[3] Solar calculations www.solarelectricityhandbook.com/index.html

[4] Solar calculations www.skyeninstruments.com

[5] Lee I E, Sim M L and Kung F W L 2009 Performance enhancement of outdoor visible-light communication system using selective combining receiver *IET Optoelectron.* **3** 30–9

[6] Lee I E, Sim M L and Kung F W L 2011 A dual-receiving visible-light communication system under time-variant non-clear sky channel for intelligent transportation system *2011 16th European Conf. on Networks and Optical Communications (NOC)* pp 153–6

[7] Islim M S and Haas H 2017 An investigation of the sunlight irradiance effect on visible light communications *2017 IEEE 28th Annual International Symposium on Personal, Indoor, and Mobile Radio Communications (PIMRC) (Oct 2017)* 8292621

[8] Khan M F F, Akif A and Haque M A 2017 Iris recognition using machine learning from smartphone captured images in visible light *2017 IEEE Int. Conf. on Telecommunications and Photonics (ICTP)* pp 33–7

[9] Sun W, Zhao C, Chen L, Li D, Bai Y, Jia W and Sun M 2015 Learning based particle filtering object tracking for visible-light systems *Int. J. Light Electron. Optics* **126** 1830–7

[10] Guo X, Shao S, Ansari N and Khreishah A 2017 Indoor localization using visible light via fusion of multiple classifiers *IEEE Photon. J.* **9**
[11] Komine T and Nakagawa M 2004 Fundamental analysis for visible-light communication system using LED lights *IEEE Trans. Consumer Electron.* **50** 100–7
[12] Keiser G 2003 *Optical Communications Essentials* (New York: McGraw-Hill)
[13] Photo diode parameters www.hamamatsu.com
[14] Chung Y H and Oh S B 2013 Efficient optical filtering for outdoor visible light communications in the presence of sunlight or artificial light *2013 Int. Symp. on Intelligent Signal Processing and Communication Systems* pp 749–52
[15] Khaligh M, Xu F, Jaafar Y and Bourennane S 2011 Double-laser differential signaling for reducing the effect of background radiation in free-space optical systems *IEEE/OSA J. Opt. Commun. Net.* **3** 145–54
[16] Karaboga N 2005 Digital IIR filter design using differential evolution algorithm *EURASIP J. Adv. Signal Process.* **2005** 1269–76
[17] Ziguang M, Wenxin W, Haiqiang J, Junming Z, Hong C, Zhen D and Yang J A novel wavelength-adjusting method in InGaN-based light-emitting diodes *Sci. Rep.* **3** 3389
[18] Farahneh H, Kamruzzaman S and Fernando X 2018 Differential receiver as a denoising scheme to improve the performance of V2V–VLC systems *ICC 2018-USA*

IOP Publishing

Visible Light Communications
Vehicular applications

Xavier Fernando and Hasan Farahneh

Appendix A

Certain VLC terminologies

The lux (symbol: *lx*) is the SI derived unit of illuminance. One lux is equal to one lumen per square meter. The lux can be converted to power density (watts/m^2) for *daylight* by multiplying it by (0.00402) [1].

The relationship between lux and the power is given by [2],

$$lx = \text{Power(W)} \times \frac{\text{LER}}{\text{Area}(m^2)} \tag{A.1}$$

Luminous efficacy (LER) expressed in lumens per watt is the rate at which a lamp is able to convert electrical power (watts) to light (lumens).

The candela (symbol: cd) is the luminous intensity in a given direction of a source that emits monochromatic radiation of frequency 540×10^{12} Hz. It has a radiant intensity in that direction of 1683 watts per steradian.

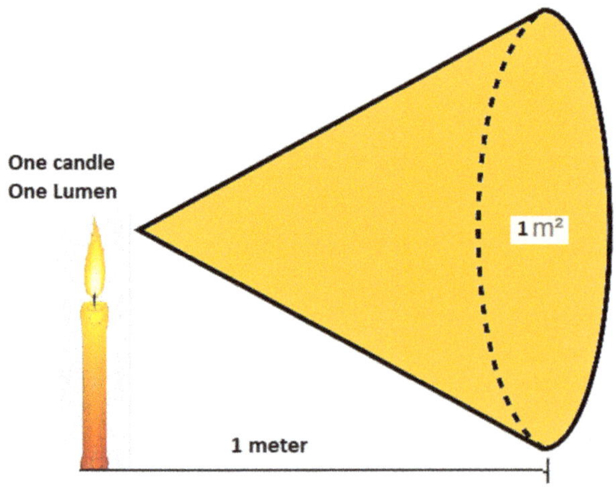

Figure A1. The relationship between the illuminance and the candela.

doi:10.1088/978-0-7503-2284-3ch7

Solar insolation is a measure of solar irradiance over of period of time—typically over the period of a single day (figure A1).

References

[1] Beshr M, Andonovic I and Hussien M 2012 The impact of sunlight on the performance of visible light communication systems over the year *Proc. SPIE* **8540** 85400F

[2] Green Business Light UK. The relationship between the illuminance and the candela. Light terminology http://www.greenbusinesslight.com

Lightning Source UK Ltd.
Milton Keynes UK
UKHW051133211219
355737UK00003B/51/P

9 780750 322829